Technology

Placing contemporary technological developments in their historical context, this book argues for the importance of law in their regulation.

Technological developments are focused upon overcoming physical and human constraints. There are no normative constraints inherent in the quest for ongoing and future technological development. In contrast, law proffers an essential normative constraint. Just because we can do something, does not mean that we should. Through the application of critical legal theory and jurisprudence to pro-actively engage with technology, this book demonstrates why legal thinking should be prioritised in emerging technological futures. This book articulates classic skills and values such as ethics and justice to ensure that future and ongoing legal engagements with socio-technological developments are tempered by legal normative constraints.

Encouraging them to foreground questions of justice and critique when thinking about law and technology, the book addresses law students and teachers, lawyers and critical thinkers concerned with the proliferation of technology in our lives.

Penny Crofts is Professor at the Faculty of Law, University of Technology Sydney.

Honni van Rijswijk is Senior Lecturer at the Faculty of Law, University of Technology Sydney.

Part of the
NEW TRAJECTORIES IN LAW
series

Series editors
Adam Gearey, Birkbeck College, University of London
Colin Perrin, Commissioning Editor, Routledge

For information about the series and details of previous and forthcoming titles, see

https://www.routledge.com/New-Trajectories-in-Law/book-series/NTL

A GlassHouse Book

Technology

New Trajectories in Law

Penny Crofts and Honni van Rijswijk

Routledge
Taylor & Francis Group
a GlassHouse Book

First published 2021
by Routledge
2 Park Square, Milton Park, Abingdon, Oxon OX14 4RN

and by Routledge
52 Vanderbilt Avenue, New York, NY 10017

a GlassHouse book

Routledge is an imprint of the Taylor & Francis Group, an informa business

British Library Cataloguing-in-Publication Data
A catalogue record for this book is available from the British Library

Library of Congress Cataloging-in-Publication Data
A catalog record has been requested for this book

ISBN: 978-0-367-23046-3 (hbk)
ISBN: 978-0-367-77137-9 (pbk)
ISBN: 978-0-429-28868-5 (ebk)

Typeset in Times NR MT Pro
by KnowledgeWorks Global Ltd.

Penny dedicates this work to Molly, Juno and Faith

Honni dedicates this work to Laura and Anika

Table of Contents

Acknowledgements

Penny Crofts' research for this book was funded by an Australian Research Council Grant Rethinking Institutional Culpability: Criminal Law, Horror and Philosophy (DE18010057).

Penny and Honni would like to thank Tina Huang for her excellent research assistance.

1 Stories of technology

The role of legal thinking in shaping techno-legal worlds

Rapid changes in technology are of urgent significance to law—and conversely, law and legal thinking are uniquely relevant to the future direction of technology. The main objective of this book is to provide a robust legal story about technological developments. To achieve this aim, we draw upon both classic jurisprudence and critical legal theory to explain, critique and advocate for law's role in responding to technology: in protecting rights, protecting people from harm, and in establishing frameworks of responsibility. We move beyond the current terrain of commentary on law and technology, which tends to focus on the (technical) purposes of specific laws, to a more general question about the purpose of law in this domain. Although "justice" is a contentious concept that is subject to debate, it is also a powerful concept, through which we analyse and evaluate law's relationship with technology. The set of questions raised by foregrounding *justice*—through the traditions of jurisprudence and critical legal studies—leads us to show what legal thinking can uniquely provide to current debates about the threats and potential of technology. Paradoxically, this doesn't mean new legal thinking *per se*, but rather going back to the basics of jurisprudence and critical legal theories and remembering what is unique about legal thinking. We need to analyse current and future socio-legal worlds that are being created through technology, noting that technology has long been part of law's work and law's domains. Throughout, we also recognise the significant shift in power that has been brought about by the increasing importance of multinational corporations to social and economic life, and the ways the corporate form is intimately connected to technology. We interrogate the legal form of the corporation and how we might shape the justice it offers.

Analysis of law's engagement with other disciplines and institutions has a long history,[1] and the engagement of law with technology is arguably one more interdisciplinary example. We argue that it is urgent that we strengthen law's story regarding technology. Technological developments are focused on overcoming physical and human constraints. In contrast, law proffers an essentially normative constraint: just because we *can* do something, does not mean that we *should*. Through the application of critical legal theory and jurisprudence to proactively engage with technological problems, this book demonstrates why legal thinking should be prioritised in emerging technological futures.

Our method draws from two established movements in critical legal theory and jurisprudence. First, a critical legal genealogy of the relationship between law and technology, highlighting law's long history of engagement with, facilitation and adjudication of, technological innovation. Second, an elucidation of theories of justice and technology, drawing on contemporary and historical thinkers. We also draw on representations of technology from both mainstream/popular and emergent/radical culture to explain key problematics and look to the ways in which popular justice thinks through the juridical implications of technology.

This chapter will consider stories of technology in law and analyse how these stories shape the possibilities of justice. Specifically, we will focus on the ways in which the dominant narratives are problematic because they downplay or even negate our ability to intervene in socio-legal worlds created through technology. In contrast, we argue that law and legal thinking should play key roles in technological innovations. This chapter provides a framework in which to situate the significance of major philosophical and jurisprudential approaches to technology. Jurisprudence and critical legal theory can contribute not only by analysing and responding to social change caused by technology, but also by shaping these emerging techno-legal worlds. Broadly speaking, jurisprudence gives us a history of and method of thinking through the emerging relationships between law and technology. Critical legal thinking gives us the means to think of the political, economic and social implications of the conceptual and material contexts of emerging technology.

Dominance of technology

There are multiple definitions of "technology"—and these definitions matter, as they impact upon the perceived potential and limitations of technology. The Oxford English Dictionary notes that there are multiple origins of the word, with Edward Phillips's definition in the

New World of Words in 1706 proffering one of the earliest sources of the word in a way that we recognise: 'a Description of Arts, especially the Mechanical'. Brian Arthur adopts a broader definition of technology, as 'the methods, practices and devices a culture uses to make things function'.[2] You can see that both these common definitions view technology as a tool and highlight the lack of constraint inherent in the concept of technology. Technology focuses on what *can* be done, rather than the normative question of whether it *should* be done. Our focus in this book is to emphasise that technology cannot be separated out from its political, economic and social context—nor from the historical and material conditions from which it arises. In other words, we discourage the idea of thinking of technology through the trope of the neutral tool that can be taken up in various contexts—rather, we encourage the reader to think of technology as essentially bound up in the conditions of its production.

We now live in a world that is permeated and saturated by technology, and the acceleration in technological developments has exponentially increased its dominance in our daily lives. We have become completely dependent upon technological systems—to eat, work, purchase, communicate and travel; with the advent of sleep trackers, the dominance of technology continues even while we are sleeping. While the various individual technologies on which we depend may appear to be separate, we can think of them constituting a vast system of interdependent technologies. Kevin Kelly has coined the term *technium* to 'encapsulate the grand totality of machines, methods and engineering processes' in and of this system.[3] The world is structured around a huge, interconnected, coordinated, incorporated system, with computers at the core. This *technium* mediates between us and everyone and everything else. The dominance of the *technium* mediates not only how we relate to each other but also, Kelly argues, patterns the structures of institutions and civilization. We are so oriented to and around technology that JM van der Laan has argued that we now have a 'technocentric culture'[4] where 'nothing else matters as much as technology'. It has been argued that the values, laws and determinations of technology begin to trump other values and meanings. For example, Allan Hanson says that legal thinking itself has been technologised. With the advent and increasing significance of legal databases in the use of law, 'lawyers are beginning to think of the law as a collection of facts and principles that can be assembled, disassembled, and reassembled in a variety of ways for different purposes'.[5] According to these views, law is one of the many major sites—alongside government, healthcare, education—that is now founded upon, constituted and predetermined by technology.

Critical legal theory: Interrogating key concepts

As we consider law's relation to technology, we should never lose sight of the significance of law's role in technologies related to slavery and colonialism—legacies that will be taken up further in Chapter 5. As just one example of the work, we should consider within the rubric of "law and technology". Ian Baucom argues that the roots of speculative finance and global capitalism can be read in the eighteenth-century case concerning the slave ship *Zong*, and the practices and institutions of which it was a part.[6] In 1781, 133 people were murdered when they were jettisoned from the *Zong.* The subsequent legal case was not a murder case but an insurance claim for loss of property. On its journey from the Cape Coast to Jamaica, encountering shortage of provisions as a result of navigational error, Captain Luke Collingwood ordered that 133 African slaves be thrown overboard. Later in court, Captain Collingwood argued that the scarcity of fresh water made this act necessary, to preserve the remainder of the human cargo, as well as the crew. Further, he expected that the loss would be covered by the ship's insurance policy. Baucom argues that the event is central to understanding not only the specific history of the trans-Atlantic slave trade, but the foundations and ethical frameworks of modern capital more widely—not as an isolated tragedy, but as part of a continuity of logic. We would add that similarly, it is essential to understanding law's complicity in responding to violence. Baucom shows that the murders and the trials that followed revealed processes of speculative finance and capital accumulation, processes that continue today. In contemporary life, finance capital is abstracted and this effect has been heightened by technology, but we should never lose site of the material histories on which they are based—and that law's role in these histories has been unbroken.

Stories of technology

Given technology's ubiquity, the types of stories we tell as a society and as legal thinkers not only describe the socio-legal world of technology, but also construct the possibilities of that world. In *Narratives of Technology,* JM van der Laan provides an overview of the mainstream stories and myths that are told about technology. Van der Laan identifies three main myths of technology – idealistic, dystopian and ambivalent—that describe and prescribe how we understand and relate to technology. We use van der Laan's taxonomy below as it provides a means of framing the key stories of the last century about social and legal responses to the exponential growth of technology.

Affirmation and technological optimism

The first category is idealistic and it champions technology. These optimists essentially express the belief that 'the machine is going to take over'—and that this is a good thing.[7] According to these accounts, technology is, at a minimum, normatively neutral. Many accounts assert that technology, particularly the internet, is normatively positive: inherently democratic, egalitarian and fair.[8] Whether normatively neutral or inherently good, technology can be harnessed for the greater good. This narrative is based on an ideal that technological development provides extraordinary power and a life without limits, inextricably linking technological developments with the inevitability of progress.[9] David Nye has explored how different idealistic accounts of technology were essential to the "making" of the United States of America.[10] On this account, technologies such as the axe, gun, mill, canal, railroad, cars and dam, represented and facilitated progress and optimism. On this account, technologies 'created' and 'made' America what it is.[11] Nye has also noted that technology is proffered as a solution to all problems, what he describes as 'a master narrative of technological amelioration'. All of humanity will be better off, even when we are replaced by robots who perform better, cheaper, more efficiently and consistently. Similarly, Van der Laan has argued that the idealistic myth places technology in the role of hero, overcoming challenges that previously only heroes could overcome—with new fabulous powers such as speed, agility, endurance and wisdom.[12] Technology is so omnipresent, omniscient and omnipotent (especially as embodied in the internet) that it is almost as though it is a divine authority. Like God, it is 'incomprehensible and impossible to master'[13]. This utopian account conveys a set of meanings and values including efficiency, progress, freedom, and power and because of its mythical status, technology cannot be significantly challenged, questioned or escaped. Essentially, these idealistic accounts give authority and legitimacy to technological development.

Within this group, theorists such as Donna Haraway[14] and Rosi Braidotti[15] assume that technology can be used for the benefit of humankind. In 1985, in her quest to construct a 'political myth for socialist-feminism',[16] Haraway famously argued in favour of post-human or trans-human forms of existence. Haraway wanted to use high-tech culture to challenge problematic epistemological and ontological dualisms such as 'self/other, mind/body, culture/nature, male/female, civilized/primitive, reality/appearance, whole/part, agent/resource, maker/made, active/passive, right/wrong, truth/illusion, total/partial, God/man'.[17] Her aim was to challenge and break down the distinctions structuring the Western self.[18] Haraway's ideas have since been developed in monster theory, which emphasises monsters as hybrids that transgress cherished dichotomies (such as the living/dead), thus challenging other foundational binaries, and hence operating as both threat and promise.[19] These ideas have been applied powerfully, with theorists drawing upon monster theory to challenge the rigidity and binary structure of the law, particularly around issues of gender and sexuality.[20] Monsters have the potential to contaminate and undermine cherished borders, to blur and weaken dividing lines that affirm binary relations. By transgressing cherished borders or rules (such as the boundary between life and death, or between human and inhuman), they challenge other cherished rules. On these accounts, admixtures of genres and borders offer political promise or a 'reverse discourse'.[21] By transgressing cherished binaries, monster theory has the potential to then challenge other binaries that structure the law, and which would otherwise seem axiomatic, with the potential for radical change. Braidotti followed in the footsteps of Haraway and argued for transforming humans through technology. She stressed that technologies are 'normatively neutral' and that there is transgressive potential in a 'merger of the human with the technological'.[22] For Braidotti, the 'becoming machine', merging the human with the technological, will combat or eliminate racism and sexism.[23]

There are a series of films that portray artificial intelligence in positive ways. For example, in *Blade Runner* (1982, Ridley Scott, Warner Bros), Harrison Ford plays a bounty hunter whose job it is to 'retire' androids, but he starts a relationship with one who is revealed to have qualities that exceed that of a machine. In *Aliens* (1986, James Cameron, 20th Century Fox), the android Ash is heroic and saves Ripley from the alien. In *Her* (2015, Spike Jonze, Warner Bros), humans and machines fall in love and the machine saves the human being demonstrating that he is capable of love. These films challenge the notion of what it means to be human, with the robots appearing to be guided by the values that we idealise as human, particularly in terms of emotions and care for others, which are frequently lacking in other human characters in those films. This genre of film is informed by an optimistic myth of technology, where technology improves or is even better than us.

This devotion to and trust of technology is reflected in law and legal education, including our dependence upon hard and software, techniques, measures, and online learning platforms.[24] These approaches adhere to the principle of social constructivism and the belief that human action shapes technology, with technology premised as either a neutral tool or a positive accentuation of the values of law. In Chapter 5, we take on the utopian dream of technology, especially as it concerns matters of race, gender and class. In particular, we look at the insights of critical legal theorists to argue that technology should not be seen as a disembodied process, but as inherently material and connected to the structural histories from which it is produced.

Pessimistic and dystopian account of technology

The counter-narrative to the idealistic myth of technology is pessimistic and dystopian. This myth aims to disrupt and challenge the idealistic myth of technological progress. Both myths assert that technology is becoming increasingly ubiquitous, but, unlike the idealistic myth, the pessimistic narrative regards technology as a threat. Despite the promise of technology, there is considerable concern about specific technological innovations, reflected, for example, in the nearly worldwide prohibition on human cloning, the ban on the use of human growth hormones in sports, debates about genetically modified foods and regulation of the use of reproductive technologies to control the sex of babies.

As we noted above, it's important to keep in mind that technology is inherently political and socioeconomic, and should not be seen as a separate tool or object. The Luddites are a good case study on this point. The word "luddite", now associated with a person opposed to new technology, was originally associated with a secret, oath-based organization of English textile workers who destroyed machinery as part of political protest based on workers' rights.[25] Although they are remembered today as hating technology, in reality their reasons for protest were more complex. The group was protesting against manufacturers who used machines in a 'fraudulent and deceitful manner' to get around standard labor practices. That is, the Luddites were not opposed to machines *per se*, but rather they wanted machines that made high quality goods and that were run by workers who had gone through apprenticeships and were paid decent wages.[26] One technology the Luddites attacked was the stocking frame that Queen Elizabeth I had initially denied a patent because it was feared that it would displace traditional hand-knitters. Their protests occurred at the start of what the Scottish essayist Thomas Carlyle called the "mechanical age". The Luddites provide an early example of the link between the state and technology through a concern for property.[27] In response to the Luddites, the British government passed emergency capital offences such as the *Frame Breaking Act* (1812)[28] and *Malicious Injuries to Property Act* (1827).[29] The Luddites were prosecuted by the British government in a mass trial, and sixty to seventy Luddites were hanged and many were transported to Australia. In his maiden speech, Lord Byron criticised the *Frame Breaking Act*, which used legal power to fulfil class rather than community interests.[30] The Luddites exercised what Eric Hobsbawm has called 'collective bargaining by riot',[31] to struggle against fundamental changes in labour relations—incipient unemployment, declining living standards and a rapid change in status by craftspeople threatened by inventions—all concerns that remain relevant today. The fact that Luddites are remembered for their destruction of, and antipathy to, machines, rather than the reasons for their actions highlights the dominance of the trope of technology as a tool (and the narrative of technology as inherently positive).

Early in the nineteenth century, Thomas Carlyle had already expressed fears that technology had come to dominate all aspects of culture and society: 'Men are grown mechanical in head and in heart, as well as in hand'.[32] The workers used technology to destroy technology—using massive sledgehammers made by a local blacksmith. While the Luddites were able to aim their sledgehammers at specific machines, in our age, tech targets are more difficult to locate—our technology is as nebulous as the cloud.

Nineteenth century political economist Karl Marx held contradictory views about technology, asserting that it 'progressively enslaved and alienated the worker, while paradoxically preparing the conditions for proletarian liberation'.[33] But Marx did consider that the very 'continuity, uniformity, regularity, order and even intensity' of mechanised labour harms the worker.[34] In *Capital* (1867), Marx devoted a section to the 'The development of machinery'—characterizing the steam engine as 'a mechanical monster whose body fills whole factories, and whose demon power, at first veiled under the slow and measured motion of his giant limbs, at length breaks out into the fast and furious whirl of his countless working organs'.[35] The worker 'becomes a mere appendage to an already existing material condition of production'.[36] Marx's conception of humans as subjugated to the monster/machine epitomises pessimistic myths of technology. Fears about technology subjugating and corrupting humans were portrayed in the silent science fiction film *Metropolis* (1927 Fritz Lang), which depicts a futuristic urban dystopia where wealthy industrialists and business magnates reign, while workers labour underground to operate the machines that power the city.

Tech theorists such as David Nye note that despite the longevity of technology, and its recent exponential developments, technology has offered little to no amelioration to longstanding human problems—it has not stopped war, established equality, enhanced human freedom, dignity or well-being.[37] Accompanied by this lack of success, despite all it promises, is the fear that technology is becoming increasingly dominant and deterministic. For example, in *Race against the machine*, Erik Brynjolfsson and Andrew McAfee have argued that the increasing use of advanced robots and self-learning algorithms may cause mass unemployment.[38] Historically, technology has been creatively disruptive, arguably creating an equal number of jobs for the jobs that it has destroyed. However, given that computers and machines are increasingly surpassing humans in performing core competencies in non-routine cognitive tasks and adaptability, it may happen in the near-future that human jobs are destroyed and not replaced. As we know all too well when our computers or the internet stop working, technology has come to exceed our understanding and oversight—it has become 'beyond our care, maintenance and radical intervention'.[39] Technology has become more autonomous and so complex that we are unable to control or master it—the *technium* 'has grown its own agenda, its own imperative, its own direction'.[40]

This fear has been expressed in relation to lawyering—that the work of lawyers will increasingly be replaced by machines. Here, it is important to distinguish between practice management (for example, document storage, diary management software, remote-access networking) and substantive legal work.[41] All professions rely on technology for practice management. Lawyers today require a certain technical proficiency if only in practice management.[42] In fact, the argument has been made that lawyering that does not significantly take advantage of technology could be considered unethical, even malpractice, because technology-assisted legal practice can yield substantively better legal results. Let's take the common litigation task of discovery as an example. When engaged in the process of discovery, lawyers may be inundated with over a million documents, and it is neither time-effective nor cost-effective for even a large counsel team to collectively review every document. This leads many practices to use specialised software to conduct a series of searches across the entire pool of documents, for counsel to review initial search results and refine. This method allows a proportionate, defensible and high-quality review of a large scale pool of documents. Software can also be used to conduct smarter research and predict legal outcomes with algorithms, using data to compile statistics on whether cases have been approved or distinguished, to identify trends, and to find the key decisions most commonly cited in judgments that favour particular parties. These tools allow research to be highly focused, even targeted—for example, the research could be aimed at a particular judge and focus on the language she uses, the judge's preference for particular cases and even suggest arguments that have a greater likelihood of success with her.[43] The use of this technology has flow-on effects regarding costs and time-saving. If these kinds of technologies can do a demonstrably better job compared to human reviewers of identifying relevant documents, one overarching question becomes whether it is competent *not* to use that software.

At the same time that lawyers may increase their productivity and accuracy, we risk abdicating our responsibility for arguably the most valuable aspect of legal thinking, which is exercising independent judgment on behalf of clients.[44] Tech in law also raises issues of confidentiality, privilege and data-protection. But these efficiencies are already being taken up—from civil discovery processes to criminal sentencing. Susskind and Susskind suggest that technologies transgress professional boundaries which means that legal professionals are no longer considered to be the sole custodians of specialist, complex knowledge and training. The stranglehold of legal practitioners on specialist knowledge is weakening and so, potentially, is the demand for traditional professions as gatekeepers of information, knowledge and practical expertise.[45] There is the potential that the nature of professional work will change from being a craft performed by human experts to a commoditised product or service available online. With de-professionalisation and segmentation comes the possibility that technological management may compromise the possibility of authentic moral action.[46] The replacement of humans with robots may improve efficiency and reliability but also eliminate human control and use, and alienate those people who may have wanted to do specific kinds of work.[47] For example, if we think about nursebots—patients may actually prefer humans to care for them, even if nursebots are more accurate and reliable *and* nursebots replace humans who may wish to be carers.

The dystopian perspective argues that we need limits and restraint and to extract ourselves from totalizing technological order. However, a great fear in the dystopian account of technology is that law is incapable of imposing those limits and restraints. Lyria Bennett Moses has critically analysed the metaphorical tendency to depict law as being in a race with technology, with law inevitably cast as the loser.[48] This is part of a broader perception that technology is speeding ahead and *nothing*, including law, is keeping up with it.[49]

Another key fear is that technology will become the measure of all things—and the only measure of value.[50] These narratives fear the subjugation of humans and our relative lack of value to machinery—'technical rationality implies a contempt of human reason'.[51] A key theorist of this counter-narrative—Lewis Mumford—fears that the myth of the machine promotes the 'mechanical world picture'—where both the natural world and symbols and values of human culture are 'cut solely to the measure of the machine'.[52] According to Mumford, this ideology privileges a denatured and dehumanised environment in which technology can flourish in the absence of any limits except those of technology itself. On this account, humanity is devalued, human ability is inferior and insufficient. The central values of technology are utility, optimization and efficiency. The fear is that rather than humans using technology, technology will increasingly use *us*. There is nothing inherently just or beneficial to society in the growth and advancement of technology.[53] In fact, many dystopians argue that technology undermines humanity and society. For example, Morozov has argued that far from delivering on the promise of the internet as free and equal, or being (at best) normatively neutral, internet centrism has meant that 'deliberation and debate are silenced; technocrats and administrators are given free reign; deeply political, life-altering issues are recast as matters of efficiency'.[54] The French sociologist Jacques Ellul argues that technology is now the 'universal social order'.[55] Technology has become the key measure of all things and the *only* measure of value. Ellul argues that technology assimilates and absorbs. He aims to dispel the myth that technology increases freedom. Rather, we only have set choices within the technological framework and we are no longer in control. Technology controls us. Our only "choice" becomes binary, those terms set by technology: to acquire, use and rely on technology for everything, or not. Film examples within this ideology include the *Terminator* films, with the underlying premise that super-intelligent machines have taken over in the future and are dangerous to humanity. The more extreme cultural example is *The Matrix*, where all human choice and agency is revealed to be an illusionary deception on the part of machines in a world where humans' only use is for their battery-power.

An interesting dystopian argument is that technological developments have the potential to not only displace law but to compromise the moral development of humans and communities. The legal theorist Roger Brownsword has made this argument in relation to the increasing use of technological management by governments and corporations to regulate society. Brownsword points to three levels of regulation. The first is classic legal regulation, such as imposing a speed limit. This form of regulation communicates on a normative or moral level—that is, you *should* do this. The second level draws upon technological developments to increase the likelihood of being caught for breaches—such as speed cameras. This level is prudential—it appeals to people's rationality and morality by increasing the risks of detection or reducing the potential reward of the offending behaviour. The third level is technological management, which renders it impossible to act in ways that violate the protected interests of others. Technological management is about what is possible and impossible.[56] An example of third level management includes cars that won't allow a person to drive if they test over the limit or fail to put on their seatbelt. In these cases, drivers will not exceed the speed limit because their vehicles are immobilised from operating. If this form of technological management replaced road traffic rules, technological management has the potential to render regulatory criminal law redundant. As Brownsword notes, the replacement of regulatory criminal law is not necessarily negative. Regulatory offences have been called 'public welfare offences'.[57] They are frequently strict or absolute liability (that is, they do not require any mental fault element), and regulate a large part of contemporary life, including public health, road safety, environment, theme park use, food safety and workplace safety. The increasing reliance upon regulatory offences has long been criticised by academics, asserting that they are not 'real crimes',[58] even though penalties attached may be very serious, because they do not establish the fault (subjective culpability) of the offender. Brownsword argues that technological management may extend beyond strict liability offences, so that the civil law area of, say, contract law may become unnecessary altogether. Of course, "smart" contracts already exist that are automated, so that if the contractual terms are breached, the contract is automatically terminated. The effects of these smart contracts could be extended so that, for example, if someone stops making payments on a car, the lender can instruct the vehicular monitoring system to not allow the car to start and to signal the location where it can be picked up.[59] Zuboti argues that this is not a new form of contract but in fact an *un-contract*—it transcends the contract form by stripping away governance and the rule of law:[60]

> Rather than enabling new contractual forms, these arrangements describe the rise of a new universal architecture somewhere between nature and God that I christen the Big Other. It is a ubiquitous networked institutional regime that records, modifies, and commodifies everyday experience from toasters to bodies, communication to thought, all with a view to establishing new pathways to monetization and profit. Big Other is the sovereign power of a near future that annihilates the freedom achieved by the rule of law.[61]

Zuboff argues that this kind of anticipatory conformity is worse than the panopticon. There are no avenues of escape. The impermissible becomes impossible.

Crucially, Brownsword argues that this kind of technological management has the potential to affect our moral capacity. The first two levels of regulatory intervention maintain the normative aspect of law—that is, although the law expresses what a person *should* do, they maintain the agency and capacity to choose whether they obey the law or not. In contrast, the third level of technological management overrides any possibility of moral choice and removes our agency. Unlike the law, which communicates on the normative level about what we should and should not do, this level of technological management removes the possibility of choosing at all. Likewise, Ian Kerr argues that 'by automating [moral choice]—by removing people from the realm of moral action altogether, [we] thereby impair … their future moral development'.[62] Kerr considers the example of digital locks which preclude people from using digital works in ways that copyright owners do not wish them to be used, so that technological management 'places coercive limits on moral actors, preventing them from acquiring access to an adequate range of life's options'. If humans are to be moral agents, they must be able to choose to be immoral, not locked out of it by technological changes'.[63] For Kerr, this kind of technological management, even where it is used for good end, 'comes at a moral price'.[64] The comparison is often made with pilots who rely so heavily on the autopilot that they lose the capacity to fly a plane. If the capacity for moral choice is removed in a pervasive way through technological management, we may lose our capacity or habit for moral reasoning. This has led Brownsword to argue that technological management 'might corrode the conditions for flourishing human agency and, especially, for moral community'.[65] Similarly, un-contracts preclude any necessity for community trust. Technological management eliminates discretion, whether by enforcers or citizens.

For example, autonomous vehicles preclude the possibility of a driver responding to dilemmas such as the trolley problem (which asks us to consider which innocent person to kill)[66] and the tunnel problem (which asks us to choose between killing a passenger in the vehicle and a child outside the vehicle).[67] Normative laws leave room for friction and tension, including the availability of an expression of responsible moral citizenship. In contrast, technological management may remove the opportunity for conscientious objection and civil disobedience. Technological management radically reduces opportunities for traditional direct acts of civil disobedience and may lead to a diminution of responsible moral citizenship. As Brownsword points out – what is most dystopian about George Orwell's *1984* and Aldous Huxley's *Brave New World* is not that human *existence* is compromised, but that human *agency* is compromised. Likewise, Anthony Burgess', *A Clockwork Orange*, portrays a disturbing world in which human moral agency is compromised. The moral philosopher Mary Midgley makes a similar argument—that in order to be free, we must have the capacity to choose between virtues and vices: 'Our nature provides for both. If it did not, we should not be free'.[68]

In addition to challenging the idea of technology as normatively neutral (at best), skeptical tech theorists point out that technology has not provided solutions we actually need, but instead created problems we can do without. For example, we initially encountered Google.com as a tool to search the internet but have subsequently found that the corporation behind the search engine has been using Google.com to mine *us* for our data, generating huge income for the company. Although technology has created jobs, it has also led to a lot of job losses, with people fearing that 'robots will replace us all'. Technology developed herbicides and pesticides to kill weeds and pests, but these toxic chemicals have in turn done terrible damage to the environment.[69] Antibiotics have saved lives but have also spawned antibiotic resistant diseases. While technology has created new, more effective forms of surveillance that assist in the prevention of crime, [70] many of these new technologies also augment structural sexism[71] and racism,[72] cement poverty, and have created new crimes.[73] The story of Frankenstein epitomizes the dystopian narrative of technology. Doctor Frankenstein undermined fundamental dichotomies of life/death and self/other in his use of science to create his monster. The story is one of scientific irresponsibility, where Doctor Frankenstein was concerned only with the question of whether he could bring a conglomerate of body pieces to life, rather than whether he should. While he may have had good intentions, his creation had unintended consequences. The examples of new problems are so ubiquitous that Ellul has argued that '[t]echnical solutions bring with them the very evils they are supposed to remedy or produce worse ones in another area'.[74]

Ambivalence about technology

While the optimistic narrative suggests we *need* do nothing about technological developments, and the dystopian narrative suggests we *can* do nothing about technological developments, the third narrative is ambivalent about technology's benefits and harms, and in turn about what we should do. It 'typically admits to problems caused by or inherent to technology, but calls for better technology which remedies or no longer causes such problems'.[75] This narrative portrays humanity's intimate alliance with technology and is often a feature of science fiction films which are ostensibly critical of technological developments. Thus, in the *Terminator* series, although evil terminators are sent back through time, "good" terminators are sent back to protect humans from machines and rely upon weaponry to defeat the villainous machines. The ambivalent narrative of these science fiction films is unsurprising, especially given that the films are using technologies in order to be made and distributed.

This approach calls for making the most of the potential of tech, and guarding against the worst. Kieran Tranter argues in his book *Living in Technical Legality* that law itself is technology, and technology, humanity and law merge into a "monstrous hybrid figure". Rather than seeing tech as a dehumanizing tool that calls for a saving law to reinstate humanity, Tranter advocates "technical legality", theories of life-affirming legal networks that co-exist with technology. We should not hope for a future that tries to separate technology from being.[76]

But the narrative of ambivalence may also lead to apathy, where it is believed that it is impossible to change the course of technology, so we may as well accept it. Accordingly, although the narrative is ambivalent, it tends to align with the dominant story of technological idealism. An example of the use of the narrative of ambivalence lies in the work of Kevin Kelly, who coined the term *technium* to express the imbrication and dominance of technology. Kelly accepts that 'most of the new problems in the world are created by previous technology',[77] and that technology 'monopolizes any activity and questions any non-technological solution as unreliable or impotent'.[78] Despite evidence that he provides in his book that the all-pervasive and encompassing technological system of the *technium* robs us of our freedom, he still believes there is net human gain from technology in the form of 'increase of freedom, choices and possibilities'.[79] Technology can expand or reduce human possibilities. We need legal stories for both.

The relationship of law with technology

There are many fears and assumptions about the relationship of law and technology, frequently informed by the defining narrative of Frankenstein – a story of scientific hubris without limits which has unintended consequences. This story is expressed particularly in relation to disruptive technology—innovations that disrupt existing regulatory or dominant schemes. It is argued that while a technological development is emerging, there is insufficient information about its potential harms and benefits. But once the (unintended) consequences become apparent, the technology is integrated into the society that uses it to the extent that it can alter the status quo and, in the process, become more resistant to regulation. This suggests that law has only a short window for effective regulation and must attempt to predict the potential future impacts of innovation. This is consistent with Lyria Bennett-Moses' critique above of the myth that law is in a race with technology and is failing to 'keep up'.[80] There is a certain amount of hysteria in this account. On a practical level, new technologies raise few concerns for law. The bulk of technological developments fit comfortably within existing legal regimes.[81] Moreover, this account belies the extent to which there are a variety of mechanisms and institutions which work together to ensure that law can adjust along with (unexpected) technological change and/or positively influence future design of particular technologies.[82] In Chapter 2 of this book we disrupt the Frankenstein's monster narrative by highlighting the long history that law has of engaging with technology through the case study of an analysis of offences of dishonesty. We also show that law has a long history of facilitating, even being complicit with, corporate technology, and that the separation of "technology" and "law" is another myth.

Bennett-Moses and Brownsword both make similar arguments about returning to first principles in law to guide the regulation of technological developments. For example, Bennett-Moses accepts that law cannot get ahead of technological developments—it is impossible to predict which socio-technological changes will occur *and* result in widespread change.[83] However, the idea of a race lets technology sets the terms of engagement. Bennett Moses has argued that rather than make a doomed attempt to predict future technology, law and policy makers and interpreters should focus on the values and purposes of specific laws. Thus 'we need to think more broadly about how to regulate to protect values and minimise harm in light of an evolving socio-technical landscape rather than simply asking how technology ought to be regulated'.[84] We argue that this ideal can be extended from focusing on the purposes of specific laws to a more general question about the purpose(s) of law. Jurisprudence offers a resource to engage with the purpose(s) of law, the nature of laws and legal systems, about the relationship of law to justice and morality and about the social nature of law. For example, although H.L.A. Hart is famous for an account which focused on a formal account of how law is to be recognised and asserted a separation of law from morality, he also accepted that law requires a 'minimum content' for the continued existence of the human race, so that we do not become a 'suicide club'.[85] This idea of law's minimum content can be drawn upon to limit destructive technological developments.[86] His secondary rules about rules of recognition and how we change law are increasingly important in ensuring that law adjusts legitimately to, and imposes restrictions upon, technological changes and challenges.

In contrast, Brownsword draws upon a natural law tradition of jurisprudence, which emphasizes law's integral relationship with something beyond law, such as nature, reason or justice. Natural law arguments became particularly salient post-World War II, where analysts argued that although the Holocaust, which killed more than five million people (particularly Jews), was enshrined in law, it was so morally wrong that citizens had no duty to obey.[87] Brownsword argues that law has three purposes. The first is to protect the 'commons' i.e. the 'preconditions for any form of human social existence, for any kind of human community, are maintained and protected; the second is to articulate and respect the distinctive fundamental values of the community; and, the third is to maintain an acceptable balance between the legitimate (but potentially competing and conflicting) interests of members of the community'.[88] The legal theorist Stephen Riley has expressed this in terms of the legitimacy of the state: 'no legal system should threaten the continued existence of the society it is intended to serve'.[89] Mireille Hildebrandt has expressed these ideals in terms of our aspirations for law:

> If we do not learn how to uphold and extend the legality that protects individual persons against arbitrary or unfair state interventions, the law will lose its hold on our imagination. It may fold back into a tool to train, discipline or influence people whose behaviours are measured and calculated to be nudged into compliance, or, the law will be replaced by techno-regulation, whether or that is labelled as law.[90]

This idea that law has a hold on our imagination is particularly captured in our beliefs in classic tenets of the law—including the Rule of Law and the belief in the relationship between law and justice. These ideals are explored in jurisprudence and applied throughout this book. The Rule of Law can be conceived in different ways, but at a minimum expresses the idea that all are subject to the rule of law, including rulers. The Rule of Law is drawn upon particularly to protect individuals against the arbitrary exercise of power by the state, and is expressed through procedural rules in criminal law such as the presumption of innocence and the requirement that the prosecution prove the guilt of an accused beyond a reasonable doubt. The Rule of Law has also been expressed as demanding responsible citizenship in terms of respect for the law.[91] There are some theorists who argue that the Rule of Law does not communicate anything about 'fundamental rights, about equality or justice',[92] but others have drawn upon the Rule of Law in a substantive way. For example, Ashworth and Zedner have explored the Rule of Law in relation to criminal law, arguing that it imposes upon the state a core duty to protect and secure its citizens a duty of justice. This duty of justice requires that the state treats people as responsible moral agents and respects their human rights and has a system of criminal justice to deal with transgressions. They express the Rule of Law thus:

> ... the norms guiding the fulfilment of these duties are contested, but core among them are liberal values of respect for the autonomy of the individual, fairness, equality, tolerance of difference and resort to coercion only where justified as a last resort ... trust ... legal articulation in requirements of reasonable suspicion, proof beyond reasonable doubt, and the presumption of innocence ... the imperative to prevent, or at least to diminish the prospect of wrongful harms thus stands in acute tension with any idealised account of a principled and parsimonious liberal criminal law.[93]

In Chapter 3, we demonstrate the power of the concept of the Rule of Law as an antidote to the misuse of Big Data by government. We also interrogate how law's relationship with corporate power and technology has in turn shaped the concept of the Rule of Law.

Another powerful ideal associated with the legal system is justice. The theorist Jurgen Habermas has made a justification for the sustained relevance of the ideal of justice, despite all the evidence of the legal system's failure to deliver justice. There is a commonality of failed aspirations between law and technology. Both have been held up as promising ideals: law holds out the promise of justice, technology holds out the promise of extending human potential. However sadly, on a practical level, it is recognised that law all too frequently fails to deliver justice, while technology has patently not delivered a freer, more equal society. We draw upon Jurgen Habermas' theory about the gap between fact and norm to emphasise the continued salience of the legal system's pursuit of justice. Habermas bases his theory on the gap between law's aspirations to be just versus the realities of the legal system.[94] Habermas contends that even though the legal system does not always achieve ideals such as justice empirically, this ideal is rightly accepted by citizens themselves as engaged participants. This means claims based on these ideals are valid, valuable and powerful, even though the legal system may not (ever) achieve these ideals. Even though the legal system may not achieve justice, we still judge the system by the idea of justice and continue to aspire to justice. In contrast, although technology has the potential to deliver greater equality or freedom and disrupt hidebound dichotomies, unlike law, it has no inherent norms and values. Technology asks whether something can be done, whilst law asks if it should be done.

An ongoing issue raised for law is whether or not new technologies can fit within existing categories or if they require new approaches. A central argument for this book is that instead of being concerned with individual technological developments, the legal system should return to first principles, which themselves may require debate in a critical and reflective way. This operates at different levels. First, this may require questions as to what values and interests law is seeking to protect. For example, autonomous vehicles may seem to challenge existing road rules, but if we return to the fundamental value that road rules are aimed at protecting—safety— then in many cases, autonomous vehicles should not require exceptional rules, as much of the evidence points to autonomous vehicles being safer than those driven by humans. In Chapter 2, we highlight that the question of values can be complicated and open to challenge, but critical engagement and debate about these values and the kind of community we aspire is vital. Second, law is frequently involved in balancing competing interests in a quest for outcomes that are just and equitable. For example, who is responsible for the costs of crashes associated with autonomous vehicles? A legislative compensation scheme needs to be created to handle autonomous car vehicle crashes which balances the interests of society in encouraging corporations to invest more in autonomous vehicles (which are safer) and those of drivers who have obeyed all instructions but where the autonomous vehicle has still crashed. In Chapter 5 we draw upon critical legal theorists as a means to critique the inherent sexism and racism of AI.

There is a concern that our legal and jurisprudential repertoire is not adequate for the challenges created by technological development. While Dworkin pointed to the integrity of the law as a guiding principle of judicial interpretation of hard cases,[95] technological developments may be so disruptive that the orthodox doctrinal questions we ask may become increasingly irrelevant. For example, Brownsword muses that technological management may require us to update our conceptions of force as the pinnacle of state intervention. Technological management is particularly enticing for regulators because, unlike legal regulation, it guarantees success (leaving aside the possibility of overriding the system or technological malfunction).[96] There is a tendency by regulators to think in a technocratic way, that is, in terms of instrumentalism—what works? But theorists such as Tamanaha have argued that instrumentalism undermine the rule of law and ideals that are fundamental to good governance:

> [T]hat the law is a principled preserver of justice, that the law serves the public good, that legal rules are binding on government officials (not only the public), and that judges must render decisions in an objective fashion based upon the law.[97]

While there might be short-term gains in completely preventing (specific) crimes, technological management reduces political and civil liberties. Brownsword argues that theorists have tended to frame the pinnacle of criminal legal intervention in terms of coercion by threat and/or force. However, we need now to re-imagine liberal values in an age of technological management, 'one of the priorities is to shake off the idea that brute force and coercive rules at the most dangerous expressions of regulatory power'.[98] The same kinds of values and laws that are applied to the justified use of criminal law should also be applied to the use of technological management. Hart's secondary rules about how to change and enforce primary (substantive) rules will become even more important in ensuring that the use of technological management is within acceptable limits.

Tech, corporations and legal thinking

Traditionally, it is argued that the legal system organises and expresses how individuals, society and the state interact with each other. For example, the rule of law is expressed as the protection of individuals from arbitrary exercise of power by the state. A central theme in this book is that we need to reshape our categories of central players and legal subjects to include corporations as key players. While corporations have long been recognised as legal subjects, the criminal legal system has great difficulty in responding to harms caused by corporations. This is because the primary legal subject is conceptualised and constructed around the human being. Technological developments raise questions about the (increasing) power of corporations because corporations are frequently the drivers, producers, suppliers and users and innovations, and not only individuals but also states are dependent on corporations for tech needs. This is particularly problematic because whilst the state is legitimated on the basis of acting in the interests of its citizenry, the central purpose of (most) corporations is profit. As we argue in Chapter 4, this question of whether or not corporations can be adequately regulated is particularly biting for tech companies due to a combination of factors. Many of these firms are oligopolies, upon which individuals and governments alike are completely dependent and yet have little to no capacity to independently remedy when issues that arise. The COVID 19 pandemic has only increased individual and governmental reliance upon tech companies. At the time of writing, Amazon became the first American company to be valued at more than $US2 trillion.[99] Artificial intelligence and automated decision-making tools are increasing in power and centrality and tech companies have huge troves of private data that can be relied upon by the state and/or sold on. These companies are at the forefront of technological innovation, and may be caught up with the factual question of what *can* be done, as opposed to the normative question, of whether it *should* be done. All these issues arise in a field in which there is little to no government regulation or intervention.[100] Combined together, companies such as Google are so new and fluid and the threats that they pose to society are so invisible, that the legal system is struggling to impose sufficient values and restrictions.[101]

The power of the tech industry is also revealed in the way that it has arranged for self-regulation as a way to evade legal and regulatory overview. For example, in the area of AI, "ethics centres" (usually funded by the tech industry) have boomed and have tended to create an extra-legal jurisdiction, using a soft ethics approach in the absence of the coercive power of the state. While ethics is important, it is essential to recognise that legal and ethical values are distinct. Law has a distinct and important role in intervening and regulating, overseeing and co-creating our socio-legal realities. As we argue in Chapter 5, ethics as applied in the current AI field is largely ineffective and prone to manipulation, especially by industry actors. The turn to ethics as a substitute for law (and legal regulation) risks its abuse and misuse.[102] Accordingly, there is an urgent need for a coherent legal approach to addressing ethics, values and moral consequences of technological developments.

Tech informing legal thinking

Tech is not a neutral and separate tool from law; nor is it separate from the material, economic and political conditions from which it arises. The complexity and inter-relation between teach and law means that technology has always shaped law and legal thinking, at the same time that law shapes tech. The critical work we need to do is across the board—rethinking law, tech, and the interests that they serve (and don't serve).

In addition, we argue that the impacts of technological developments need to be conceptualised not only in terms of the individual, but also the community or society. Much of classic legal doctrine is conceptualised in terms of the impact of the individual. However, as we argue throughout the book, this focus on the individual is inadequate to capture the impact of technologies. One technique that we suggest is that while a specific domain of legal doctrine may be exhausted by technological development, it is possible to turn to other legal domains to conceptualise the challenges of technological development. For example, in Chapter 3 we argue that analyzing data collection in terms individual impact does not reflect the wide-scale impact of collection. We turn to environmental law and classic tenets of criminal law to explore data collection in terms of pollution and public harms.

In summary, our position is that we don't yet have an adequate conceptual and critical language for the rapid mediation of the world through technology. It is difficult to write simple stories about the technological world. Our intervention here will be to articulate alternate stories of law and technology, myths that place justice, critical legal thinking and even the Rule of Law at their centre, and also the agency of lawyers and critical legal thinkers to take the lead on how technology might be shaped. This book aims to disrupt the techno-centric narrative and provide a more law-centric narrative that is concerned with humans, life, biology, society and the environment.

Notes

1. Goodrich has analysed how law engages with other disciplines such as medicine through the concept of interdiscourse. Peter Goodrich, *Legal Discourse: Studies in Linguistics, Rhetoric and Legal Analysis* (St Martin's Press 1987).
2. Brian Arthur, *The Nature of Technology: What It Is and How It Evolves* (Free Press 2009). 27.
3. Kevin Kelly, *What Technology Wants* (Viking 2010). 12.
4. JM van der Laan, *Narratives of Technology* (Palgrave MacMillan 2016). 13.
5. Allan Hanson, *Technology and Cultural Tectonics: Shifting Values and Meaning* (Palgrave Macmillan 2013). 131.
6. Ian Baucom, *Specters of the Atlantic: Finance Capital, Slavery and the Philosophy of* History (Duke University Press 2005).
7. Ibid 52. Quoting Clarke at 229.
8. Steven Johnson, *Future Perfect: The Case for Progress in a Networked Age* (Riverhead Books 2012).
9. Howard Segal, *Technological Utopianism in American Culture* (University of Chicago Press 1985). 1.
10. David Nye, *America as Second Creation: Technology and Narratives of New Beginnings* (MIT Press 2003). 18.
11. ibid. 292.
12. Many of the super hero franchise films portray people (usually men) who harness technology to become heroes. Ongoing examples of the genre include Batman and Ironman, the Bionic Man is an early television series example. James Bond is an early example whose heroism is linked particularly to his access to innovative technologies and fast cars. Van der Laan is arguing that in optimistic accounts of technological development it is technology that is the hero.
13. van der Laan (n 4). 71.
14. Donna Haraway, 'A Cyborg Manifesto: Science, Technology and Socialist Feminism in the Late Twentieth Century' in Donna Haraway (ed), *Simians, cyborgs and women: The reinventions of nature* (Routledge 1991). 149–181.
15. Rosi Braidotti, *The Posthuman* (Polity Press 2013).
16. Haraway (n 14). 157.
17. ibid. 177.
18. ibid. 174.
19. Penny Crofts, 'Don't Blink: Monstrous Justice and the Weeping Angels of Doctor Who' in William MacNeil, Timothy Peters and Karen Crawley (eds), *Envisioning Justice* (Routledge 2018).
20. Judith Halberstam, *Skin Shows: Gothic Horror and the Technology of Monsters* (Duke University Press 1995); Alex Sharpe, *Foucault's Monsters and the Challenge of Law* (Routledge 2010).
21. Penny Crofts and Honni van Rijswijk, 'Traumatic Origins in Hart and Ringu' in Ashley Pearson, Thomas Giddens and Kieran Tranter (eds), *Law and Justice in Japanese Popular Culture* (Routledge 2018).
22. Braidotti (n 15). 92.

23. ibid. 97. The American adventure comedy film, *Jumanji: Welcome to the jungle* (Jake Kasdan, Columbia Pictures 2017) provides an example of technological optimism in popular culture. In this film, gender and personality are marked as fungible once the characters enter the board game via the video game system—for example, the shallow Bethany is transformed into a geeky male palaeontologist Professor Oberon.
24. There are too many examples to mention, but articles analysing the use of technology in legal education include: Eola Barnett and Linda McKeown, 'The student behind the avatar: using Second Life (virtual world) for legal advocacy skills development and assessment for external students—a critical evaluation' (2012) 8(2) *Journal of Commonwealth Law and Legal Education* 41; Desmond A. Butler, 'Technology: New Horizons in Law Teaching' in Sally Kift, Michelle Sampson, Jill Cowley and Penelope Watson (ed), *Excellence and Innovation in Legal Education* (LexisNexis 2011) 460-459; Julian Hermida, 'Teaching Criminal Law in a Visually and Technology Oriented Culture: A Visual Pedagogy Approach' (2006) 16((1&2)) *Legal Education Review* 153; Dan Jackson, 'Human-centered legal tech: integrating design in legal education' (2016) 50(1) *The Law Teacher* 82; Elizabeth Seul-gi and Anneka Ferguson Lee, 'The development of the virtual educational space: how transactional online teaching can prepare today's law graduates for today's virtual age' (2015) 6(1) *European Journal of Law and Technology*; Paul Maharg, 'Editorial: Learning/Technology' (2016) 50(1) *The Law Teacher* 15; Paul Maharg and Emma Nicol, 'Simulation and technology in legal education: a systematic review and future research programme' in Caroline Strevens, Richard Grimes & Edward Phillips (ed), *Legal Education: Simulation in Theory and Practice* (Ashgate 2014) 17–42; Mitchell Travis, 'Teaching Professional Ethics through Popular Culture' (2016) 50(2) *The Law Teacher* 147.
25. Richard Conniff, 'What the Luddites Really Fought Against' *Smithsonian Magazine* (March 2011) <https://www.smithsonianmag.com/history/what-the-luddites-really-fought-against-264412/>. Viewed 6 January 2021
26. Kevin Binfield (Ed) *Writings of the Luddites.*
27. John Locke, '*The Second Treatise of Government*' (1690) Peter Laslett (Ed). (Cambridge University Press 1988).
28. http://ludditebicentenary.blogspot.com/2012/03/20th-march-1812-1812-frame-breaking-act.html. Viewed 6 January 2021.
29. van der Laan (n 4).
30. Quoted in Karly Walters, 'Law, "terror" and the Frame-Breaking Act' (2004). 24.
31. Eric Hobsbawm, *'The Machine Breakers' and Labouring Men: Studies in the History of Labour* (Weidenfeld and Nicolson 1968). 58.
32. Thomas Carlyle, 'Signs of the Times' (1829) 49 Edinburgh Review 439.
33. Carl Mitcham and Timothy Casey, 'Toward an Archeology of the Philosophy of Technology' in Mark Greenberg and Lance Schachterle (eds), *Literature and Technology* (Lehigh University Press 1992). 78.
34. Karl Marx, *Capital: A Critique of the Political Economy: Volume 1,* vol 35 (Alexander Chepurenko ed, Samuel Moore and Edward Aveling trs, International Publishers 1996). 350.

35. ibid. 384–5.
36. ibid. 79.
37. David Nye, *Technology matters: Questions to live with* (MIT Press 2006). See also, Langdon Winner, *Autonomous Technology: Technics-out-of-Control as a theme in political thought* (MIT Press 1977); Margaret Thornton, 'The Flexible Cyborg: Work-Life Balance in Legal Practice' (2016) 38(1) *Sydney Law Review* 1.
38. Brynjolfsson, E. and A. McAfee (2011), Race Against the Machine. How the Digital Revolution is Accelerating Innovation, Driving Productivity, and Irreversibly Transforming Employment and the Economy, Digital Frontier Press, Lexington, Massachusetts, accessed 7 August 2018; http://b1ca250e5ed661ccf2f1-da4c182123f5956a3d22aa43eb816232.r10.cf1.rackcdn.com/contentItem-5422867-40675649-ew37tmdujwhnj-or.pdf
39. Albert Borgman, *Technology and the Character of Contemporary Life: A Philosophical Inquiry* (University of Chicago Press 1984). 113.
40. Kelly (n 3). 186.
41. Kevin Smith, 'Welcome to the Machine: Considering the Ethics of Legal Technology' (2019) 77 Advocate (Vancouver Bar Association) 43.
42. Michael Murphy, 'Just and Speedy: On Civil Discovery Sanctions for Luddite Lawyers' (2017) 25 George Mason Law Review 36.
43. Smith (n 41).
44. The television series *The Good Fight* (CBS 2017) portrays a contemporary American law firm that is grappling with the impact of technology. The lawyers deal with cases involving fake news, artificial insemination, cyber harassment, malfunctioning voting machines and social media. However, the impact of automation is demonstrated more critically in Episode Two (Series One), in which Boseman is initially unable to get funding to cover a class action case because they are unlikely to succeed due to the judge. Boseman responds by asking if the algorithm takes into account that he is a good lawyer, but only succeeds in getting funding by claiming that he can get the judge changed. The algorithm calculates the likelihood a lawsuit will succeed, the judge's past decisions, and the size of the judge's caseload (to provide an indication of timing).
45. Daniel and Richard Susskind Susskind, *The Future of the Professions: How Technology Will Transform the Work of Human Experts* (Oxford University Press 2015).
46. Nigel Stobbs, Dan Hunter and Mirko Bagaric, 'Can Sentencing Be Enhanced by the Use of Artificial Intelligence?' (2017) 41 *Criminal Law Journal* 261.
47. Roger Brownsword, *Law, Technology and Society* (Routledge 2019). 15.
48. Lyria Bennett Moses, 'Recurring Dilemmas: The Law's Race to Keep Up with Technological Change' (2007) 7 University of Illinois *Journal of Law* 239. Moses cites a series of examples including *See, e.g.,* Mount Isa Mines Ltd. v. Pusey, 125 C.L.R. 383, 395 (Austl. 1970) (per Windeyer J.) ("Law, marching with medicine but in the rear and limping a little"); Michael Kirby, *Medical Technology and New Frontiers of Family Law,* 1 AUST. J. FAM. L. 196, 212 (1987) ("The hare of science and technology lurches ahead. The tortoise of the law ambles slowly behind."); Rev. John H. Pearson CSC,

Regulation in the Face of Technological Advance: Who Makes These Calls Anyway?, 13 N.D. J. OF L., ETHICS & PUB. POL'Y 1, 1 (1999) ("It has become commonplace to note that these dizzying changes in science and technology can easily outstrip those systems by which we humans make critical decisions about what can and should be done by those who are responsible members of society and about how to protect those responsible members of society from those who are not so responsible."); Grant Gilmore, *The Ages of American Law* 65 ("Rapid technological change unsettles the law quite as much as it unsettles people."). See for example, Kayleen Manwaring, 'Will Emerging Information Technologies Outpace Consumer Protection Law?—The Case of Digital Consumer Manipulation' (2018) 26 *Competition & Consumer Law Journal* 141.

49. Lyria Bennett Moses, 'Agents of Change: How the Law "Copes" with Technological Change' (2011) 20(4) *Griffith Law Review* 763. Citing Popper 2003.
50. Ibid. 86.
51. Friedrich Georg Junger, *The Failure of Technology* (Gateway Editions 1956). 141.
52. Lewis Mumford, *The Myth of the Machine: The Pentagon of Power*, vol Two (Harcourt Brace Jovanovich 1970). 24.
53. The recent call by scientists for an international treaty against the development of autonomous warfare is an example of fears that the ambitions of technology are not necessarily positive for the human race. Mateusz Piatkowski, 'Fully Autonomous Weaspons Systems and the Principles of International Humanitarian Law' (Paper presented at the 5th International Conference of PhD students and young researchers: How deep is your law?, Lithuania, April 2017) <https://papers.ssrn.com/sol3/papers.cfm?abstract_id=3006230>.
54. Evgeny Morozov, *To Save Everything, Click Here: The Folly of Technological Solutionism* (PublicAffairs 2013). 134.
55. Jacques Ellul, *The Technological Society* (John Wilkinson tr, Knopf 1964). 419.
56. Brownsword (n 47).
57. Francis Sayre, 'Public Welfare Offences' (1933) 33 Columbia Law Review 55.
58. See e.g. the Canadian Supreme Court case of *R v Sault Ste Marie* [1978] 2 SCR 1299 per Dickson J at 1302-1303.
59. HR Varion, 'Computer Mediated Transactions' (2010) 100 American Economic Review 1.
60. Shoshana Zuboff, 'Big Other: Surveillance Capitalism and the Prospects of an Information Civilization' (2015) 30 *Journal of Information Technology* 75.
61. ibid. 81.
62. Ian Kerr, 'Digital Locks and the Automation of Virtue' in Michael Geist (ed), *From 'radical extremism' to 'balanced copyright': Canadian copyright and the digital agenda* (Irwin Law 2010). 247.
63. ibid. 303.
64. Brownsword (n 47). 71.
65. ibid. 87.

66. Philippa Foot, *Virtues and Vices and Other Essays in Moral Philosophy* (1978).
67. Brownsword (n 47). 16.
68. Mary Midgley, *Wickedness: A Philosophical Essay* (Routledge 1984). 3.
69. Rachel Carson, *Silent Spring* (Houghton Mifflin 1962).
70. Ronald Bailey, 'Can Algorithms Run Things Better Than Humans?' (2019) 50 *Reason; Los Angeles* 20.
71. Alison Adam, *Artificial Knowing: Gender and the Thinking Machine* (Routledge 1998); Alison Adam, 'A Feminist Critique of Artificial Intelligence' (1995) 2 *European Journal of Women's Studies* 355.
72. Sarah E Chinn, *Technology and the Logic of American Racism: A Cultural History of the Body As Evidence* (Bloomsbury Publishing Plc 2000) <http://ebookcentral.proquest.com/lib/uts/detail.action?docID=436311> accessed 14 August 2019.
73. Akintunde Abidemi Adebayo and Alaba Ibironke Kekere, 'Electronic Commerce in Nigeria: The Exigency of Combatting Cyber Frauds and Insecurity' (2016) 47 *Journal of Law*, Policy and Globalization; Mubaraz Ahmed and Fred Lloyd George, 'A War of Keywords: How Extremists Are Exploiting the Internet and What to Do about It' (Tony Blair Institute for Global Change 2017); Payal Arora and Laura Scheiber, 'Slumdog Romance: Facebook Love and Digital Privacy at the Margins' (2017) 39 Media, Culture & Society 408.
74. Jacques Ellul, *The Technological Bluff* (Geoffrey Bromiley tr, Eerdmans 1990). 93.
75. van der Laan, above n 7.
76. Kieran Tranter, *Living in Technical Legality* (Edinburgh University Press 2018). 2.
77. Kelly (n 3). 192.
78. ibid. 193.
79. ibid. 237.
80. Lyria Bennett Moses, 'Agents of Change: How the Law "Copes" with Technological Change' (2011) 20 Griffith Law Review 763.
81. For example, the horror/thriller *The Invisible Man* (2020 Blumhouse Productions) depicts a new technology of a suit that renders the wearer invisible. Whilst this suit is radically new, the crazed scientist uses it to stalk and abuse his ex-girlfriend. The actions of the scientist and the way his victim is regarded by her friends and the police is consistent with long-term domestic violence research, despite the use of the suit. Classic domestic violence legislation could be drawn upon to regulate this (mis) use of the suit. The suit may exacerbate harms of domestic violence and make it more difficult to police, but the perpetrator's actions are comprehensible (and reprehensible) within existing legislative regimes.
82. Bennett Moses, 'Agents of Change: How the Law "Copes" with Technological Change' (n 80).
83. For example, at this point in time, it would be ridiculous for laws to be made regulating the use of invisible cloaks. However, invisible cloaks may become a reality *and* widespread in the future. http://edition.cnn.com/2016/07/20/health/invisibility-cloaks-research/index.html.
84. Bennett Moses, 'Agents of Change: How the Law "Copes" with Technological Change', above n 18. 787.

85. HLA Hart, *The Concept of Law* (1961).
86. An example of this is the movement to prevent lethal autonomous weapons using human rights arguments expressed in the Montreal Declaration for the Responsible Development of Artificial Intelligence 2018.
87. For jurisprudential engagement with these arguments see HLA Hart, 'Positivism and the Separation of Law and Morals' (1958) 71 Harvard Law Review 593; Lon Fuller, 'Positivism and a Fidelity to Law: A Reply to Professor Hart' (1958) 71 Harvard Law Review 661; Desmond Manderson, 'Apocryphal Jurisprudence' (2001) 27 *Journal of Legal Philosophy* 27.
88. Brownsword (n 47). 100.
89. Stephen Riley, 'Architectures of Intergenerational Justice: Human Dignity, International Law, and Duties to Future Generations' (2016) 15 *Journal of Human Rights* 272.
90. Mireille Hildebrandt, *Smart Technologies and the End(s) of Law* (Edward Elgar 2015). 17.
91. Joseph Raz, 'The Rule of Law and Its Virtues' (1977) 93 Law Quarterly Review 195; David Dyzenhaus, 'Recrafting the Rule of Law' in David Dyzenhaus (ed), *Recrafting the Rule of Law* (Hart 1999).
92. Raz (n 91). 198.
93. Andrew Ashworth and Lucia Zedner, *Preventive Justice* (Oxford University Press 2014). 251–252.
94. Jurgen Habermas, *Between Facts and Norms: Contribution to a Discourse Theory of Law and Democracy* (1996).
95. Ronald Dworkin, *Law's Empire* (Harper Collins 1986).
96. Brian Tamanaha, *Law as a Means to an End* (Cambridge University Press 2006).
97. ibid. 249.
98. Brownsword (n 47). 172.
99. https://www.abc.net.au/news/2020-08-20/apples-market-value-tops-$us2-trillion-record/12576384.
100. See, for example, Carolyn Abbott, 'Bridging the Gap—Non-State Actors and the Challenges of Regulating New Technology' (2012) 39 *Journal of Law and Society* 329; Lyria Bennett Moses, 'How to Think about Law, Regulation and Technology: Problems with "Technology" as a Regulatory Target' (2013) 5 Law, Innovation and Technology 1.
101. Evgeny Morozov, 'Don't Be Evil' [2011] *The New Republic* <https://hci.stanford.edu/courses/cs047n/readings/morozov-google-evil.pdf>.
102. Anais Resseguier and Rowena Rodrigues, "AI ethics should not remain toothless: A call bring back the teeth of ethics," *Big Data and Society*, July 22, 2020.

2 Historical examples of law and technology

Civil law, criminal law, slavery and colonialism

Historically, law has had various relationships with the social, political and economic effects of technology: sometimes, it has sought to limit or prevent harm; and sometimes, law has been complicit in assisting corporations and individuals from being held accountable for harm. Below, we consider two case studies—the historical connection of tort's reading of value to technological change and then the history of dishonesty offences—to show that there is still dispute about the types of values and/or interests that the law is seeking to protect.

Case study: Negligence, technological change and value

Negligence law was essentially a creature of the nineteenth century, and its development was intimately connected with technological change. Injuries caused by industrialisation introduced new cases to the courts, which were initially unsure how to respond, since their main preoccupation in tort had previously been "occasional assaults, defamations, boundary disputes, noxious neighbours, trespassing cattle and the like".[1] According to John Fleming, "Parliament was otherwise preoccupied", and could not deal with these injuries, so the courts were forced to respond: "by and large, their response was to try and contain the flood by raising the barriers to recovery".[2]

Culturally and politically, the nineteenth century saw the rise of principles of individualism and self-responsibility, accompanied by a laissez faire spirit of government towards business.[3] These pressures shaped the nature of liability for injury, insuring that the cost of injury lay with the individual, not with industry or government:

> It was felt to be in the better interest of an advancing economy to subordinate the security of individuals, who happened to become casualties of the new machine age, rather than fetter enterprise

> by loading it with the costs of "inevitable" accidents. Liability for faultless causation was feared to impede progress because it gave the individual no opportunity for avoiding liability ....[4]

The strongest form of this argument was articulated by Morton J. Horwitz in his influential book *The Transformation of American Law*, where he argued negligence law was essentially invented in the nineteenth century not to compensate victims but rather to subsidise industry as fault-based liability replaced strict liability with the effect of narrowing the field of capitalist obligation:

> The law of negligence became a leading means by which the dynamic and growing forces in American society were able to challenge and eventually overwhelm the weak and relatively powerless segments of the American economy. After 1840 the principle that one could not be held liable for socially useful activity exercised with due care became a commonplace of American law.[5]

There is, however, evidence that court decisions were more nuanced than this: some judges and juries also resisted the excesses of capitalism and used tort law as an occasion to protest these excesses. Lunney and Oliphant argue that opinions such as Horwitz's fail to take account of the wide range of judicial responses, especially in the British context, the anti-capitalist approach of those judges who were members of the landed gentry.[6] In the case of railway injuries, in particular, juries often awarded large sums to plaintiffs, not only to compensate plaintiffs for their suffering, but also out of a kind of revenge against corporations, by way of protest against the ethic of the bottom line, and out of popular terror of railway accidents.[7] In 1889, a commentator wrote that most jurors had an "innate prejudice against all corporations or capitalists".[8]

Despite the ambivalence of courts concerning the injuries of industrialization, the essential value of both the contemporary moment and the future of tort were defined as *economic* in a seminal work written at the end of the nineteenth century. American jurist Oliver Wendell Holmes Jr's book *The Common Law* (1881) was "arguably the first 'modern' effort to unravel fundamental problems of the common law, including tort law, in basic philosophic terms".[9] Holmes was not only one of the most important commentators on modern American law, but also a Supreme Court judge, who created the law he described. One of Holmes's main rhetorical strategies in this work was to set late nineteenth-century values in opposition to those of the past. In olden times, Holmes wrote, the way in which the law valued suffering was straightforward: then, the value was clearly moral and was based on the uncomplicated and satisfying principle of revenge. Writing sixty or so years after negligence had transformed tort law, Holmes Jr looked back to that earlier time:

> The various forms of liability known to modern law spring from the common ground of revenge ... [I]n the criminal law and the law of torts it is of first importance. It shows that they have started from a moral basis, from the thought that some-one was to blame.[10]

Holmes used this historical argument to distinguish the late-nineteenth-century values of the common law from what preceded it and to develop his normative thesis, which was that moral values needed to be carefully delineated and separated from legal values. Behind this sentiment was a sympathy for economic value and a belief that law was a "business with well understood limits".[11] Not only text of the law, but the very minds and souls of lawyers needed to change. In 1897, he wrote:

> For the rational study of the law the black-letter man may be the man of the present, but the man of the future is the man of statis-tics and the master of economics.[12]

The first major movement in contemporary tort theory, the law and economics movement, occurred some sixty years after Holmes's *Common Law*, but was sympathetic with his approach.[13] The move-ment dominated the legal theory of tort until the 1980s, when some jurists became sufficiently outraged that legal theory had been reduced to economics to found a philosophy of tort movement that was based on the importance of moral value. These critics argued that, with the law and economics movement, "the law ha[d] 'gone a-whoring after false gods".[14]

The stories we tell about how law has responded to technological change matter. They set in motion not only how we *describe* law's rela-tionship, but how we *prescribe* it—for lawyers, law students and courts. The moral account of tort's value is written in the language of "duty of care", of neighborliness, responsibility and civil society. David Owen calls moral responsibility the "basic cement" of negligence;[15] Peter Cane argues negligence law is *essentially* an ethical system of responsibility.[16] This obligation arising in tort law is also seen to be unique in law:

> It is not the outcome of an agreement founded on self-interest, like a contract. It is not a duty owed to the community as a whole and acted on by the state, like criminal law. It describes a *personal* responsibility we owe to others which has been placed upon us without our consent. It is a kind of debt that each of us owes to others although we never consciously accrued it. Thus it raises, in a distinctly personal way, one of the oldest questions of law itself: "Am I my brother's keeper?"[17]

Theorists who base tort law's value in the moral domain acknowledge the importance of industrialisation and technological change to the development of tort law, but regard economics as being part of social value, rather than being an end in itself. For these thinkers, the change in the law's values noted by Holmes, from an ethic of revenge to one of compensation, is seen as effecting a change in relationship—from the relationship between an individual and the state to a relationship between private actors—rather than marking the supremacy of economic value. Originally, compensation was provided not to reduce the plaintiff's suffering, but to calm any plans of retaliation against the defendant and so to keep the king's peace.[18] As between the parties, damages were awarded "for the benefit of the wrongdoer's soul rather than of the victim's pocket".[19] But with the development of negligence law, the rationale of compensation changed and damage, rather than conduct, became the focus. In negligence, the defendant's conduct came to matter not because of his soul, or because of the king's peace, but in relation to somebody else's suffering. Tort law has been thought of as a kind of moral praxis—as the practical, textured compliment to philosophical abstraction. Peter Cane argues that:

> … by reason of law's institutional resources, the legal "version" of responsibility has a richness of detail lacking in the moral "version" of responsibility. Because law is underwritten by the coercive power of the state, courts cannot leave disputes about responsibility … unresolved. … Morality can afford to be vague and indeterminate to an extent that law cannot.[20]

Further, Cane argues that "outside of the law, there are relatively few norm-enforcing institutions in civil society".[21]

What is the nature of this moral responsibility and upon what is it based? The civil sphere is usually referred to in liberal theories to indicate the sphere that is separate from both the state and market. The literature on civil society is enormous and it is a key idea in Enlightenment political theories by John Locke, Adam Smith and Immanuel Kant. It was revamped by Karl Marx and has more recently been critiqued by feminist theorists, post-colonial theorists and queer theorists, who draw attention to the borders and exclusions of the civil domain. In the late twentieth century it enjoyed new life through the philosophy of Jürgen Habermas. For Habermas, civil society has political potential, existing as it does where the private and public spheres of the lifeworld meet, and consisting of:

> … more or less spontaneously emergent associations, organizations and movements that, attuned to how societal problems resonate in the private life spheres, distil and transmit such reactions in amplified form to the public sphere.[22]

The attraction of the civil sphere arises in part because it is seen to address the problem of balancing individual and social interests as "an ethical ideal of the social order, one that, if not overcomes, at least harmonises, the conflicting demands of individual interest and social good".[23]

According to Desmond Manderson, it is the neighbor principle itself that exceeds the language and principles of tort law: a "spirit" of responsibility hovers outside the law, and is constantly referred to by the law, but cannot be reduced to legal rules.[24] This responsibility has never been realised by the law and nor has tort "been entirely comfortable" with this haunting.[25]

Social and economic forces have historically carried great weight in tort's adjudication of suffering, at tension with moral value. At the turn of this century, in Australia, as in other common law jurisdictions, courts transformed tort law again and "self-responsibility" became a key term in justifying the dominance of such social and economic interests, leading to legislative reforms that limited a number of forms of liability for negligence. The purported occasion for these changes was a sense that "fun has been cancelled"—there were media reports that the greedy demands of careless, injured plaintiffs had led to impossibly high insurance premiums and so the cancellation of everything from country fairs to children's playgrounds. In fact, technological change leading to the transformation of the insurance market, interest rates, trends in global trade and competition were in the end, more influential than the number of claims made by plaintiffs in affecting the insurance industry.[26] Tort has returned to the individualist perspective that influenced the development of fault-based liability in the nineteenth century.

The term "self-responsibility" carries both moral and economic meaning: it implies a desired value of self-reliance and independence; but at the same time, it implies that society and the economy will not be responsible for the individual's suffering. The deployment of the term also has political effects, as it is an ethic that complements the contemporary dominance of neoliberal principles, which favour free-market capitalism and a minimalist role for the state.

Case study: Criminal law, dishonesty and technology

One of the first stories of technology is the myth of Prometheus, who defied Zeus and stole fire to give it to humanity. Fire was a primal technology, from which other technologies flowed. As Aeschylus wrote, "every art possessed by man comes from Prometheus".[27] As a consequence of this theft, Zeus punished Prometheus by chaining him to a rock. Each day an eagle came to eat his liver, which would grow back only to be devoured again. In addition, Zeus also punished humanity by creating Pandora, who released from her jar "the price for the blessing" of fire, that is, technology, all the plagues upon humankind.[28]

This is a founding story where technology comes with the intention to benefit and improve humanity, but is accompanied by a price, a curse. It is also a story of law, that of theft, transgression and punishment. Theft and related dishonest acquisition offences provide an example of long-term legal engagement with socio-technological change. Dishonest acquisition offences are of interest because there are so many different ways to dishonestly take so many different things, and technological developments have been part of the reason why methods and things have changed across time and place. Legal systems have tended to respond to the new ways of theft case by case, resulting in a confusing and complex web of laws and this has been exacerbated by technological development. The common law offence of larceny dates from the thirteenth century, and needless to say, twenty-first century technological developments were not foreseen when these early offences were created. Metamorphosis in the offence of larceny and its offshoots has occurred across the two axes of ways of stealing and what types of property could be stolen. The law has responded to technological changes in differing ways, with different levels of success. As we argue below, law has a counter-intuitive concept of (the technology of) money, which has created difficulties in dishonesty offences across time, exacerbated by electronic money. One ongoing question is whether ways of acquiring property dishonestly can and should fit into existing legal categories, or whether these methods are sufficiently different that new legal categories need to be constructed. An issue that has also been raised with increasing urgency is the challenge of balancing corporate and state interests against those of individuals. We argue that while the law has responded with some ease and agility to new forms of dishonesty by individuals, the law has not been very successful in engaging with corporate dishonesty.

One of the key arguments in this book is that the law needs to return to fundamentals, to consider what interest or value it is seeking to protect. However, dishonest acquisition offences are aimed at protecting multiple interests and values, and these have changed across time. In his book, *Thirteen Ways to Steal a Bicycle*, Stuart Green points to the "messy complexity" that currently characterises the law of theft.[29] The contemporary perception of dishonest acquisition offences is that their central aim is to protect private interests of property and this is the reason why they tend to be categorised as property offences. However, this perception is a product of the Enlightenment, particularly from Blackstone and his efforts to find reason in the structure of the common law. Earlier conceptions of why dishonest acquisition is sufficiently blameworthy continue to underlie contemporary criminal law. The offence of larceny continues to form the basic theft offence in some jurisdictions, while others have replaced the offence (and those of embezzlement and fraud) with the all-encompassing offence of 'theft'.[30] The legal theorist George Fletcher has argued persuasively that initially the offence was based upon the underlying pattern of blameworthiness of manifest criminality, that is the idea that one could recognise a thief based on a collective image of acting like a thief.[31] Thieves could be seen thieving, they could be caught in the act. A thief would be recognisable as someone who was in the wrong place at the wrong time. A thief broke the close and endangered the security of the manor, trespassing where the thief was not meant to be. There were a multitude of interests jeopardized by this invasion including the security and well-being of the community. According to Roman and Biblical law, the manifest thief caught in the act 'red handed' was subject to execution on the spot, leading Fletcher to argue that the thief in the night was regarded as the paradigmatic threat to the community, the wrongdoer who should be killed on the spot. On this account, dishonest acquisition was a crime that disturbed and undermined the community. A common theme of early law was that the offence was aimed at dishonest immoral behavior.[32]

However, Fletcher argues that with the industrial revolution, this pattern of blameworthiness no longer served the community. The industrial revolution was a result of massive socio-technological change. People no longer knew everyone in their community. They traded with strangers who they never saw again. These strangers might appear trustworthy, but they could use trickery to take advantage of them. In order to succeed at theft, they needed to not look like a thief, that is, not *seem* manifestly criminal. Fletcher argues that in response to this shift in social relations, property offences were transformed from a focus on manifest criminality to subjective culpability. The courts no longer focused on whether a person looked or acted like a thief, but upon the accused's culpable intent. Rather than overturn the earlier cases the courts reinterpreted them, asserting that earlier cases had been decided on the basis of felonious intent. As a consequence, this area of law is extremely complex. This is an example of legal change in response to socio-technological developments. The changes were accomplished by a reinterpretation of existing categories based upon an underlying paradigm shift. Accordingly, on the face of it the law remained the same, but it was reinterpreted and applied to reflect developing community needs.

The offence of larceny retained very specific requirements based on its earlier conception of the thief in the night, including a requirement of physical movement of property (asportation), trespass and lack of consent. But with technological developments these specific requirements became difficult to meet. The common law of larceny protected physical things. Can intangible property or rights be stolen?[33] How can intangible property move? Trespass requires an offence against the possession of another, but there are many different ways of possessing things. The offence of larceny required a lack of consent by the person in possession, but what about consent obtained by trickery or fraud? And can machines consent? In response to these challenges, some jurisdictions overturned the offence of larceny and introduced a new offence of theft. The offence of theft covers intangible property, removing the key requirements of larceny of property belonging to another and asportation—and sought to overcome supposedly archaic distinctions between offences of larceny, embezzlement and fraud to create a catch-all offence of 'theft'.[34] Many jurisdictions have introduced specific offences to deal with specific challenges as they arise, resulting in a hodge-podge of laws. Despite these reforms, the challenges of the myriad of methods and objects of dishonest acquisition remain.

We cannot hope to cover all dishonesty offences, but the aim of this part of the chapter is to provide different examples of challenges that have arisen as a result of technological developments and assess how the law has responded. We highlight the choices made in law—can and should technological developments slot into existing legal categories, or should new legal categories be created? What kind of interests and/ or values are being protected? There are not always easy answers to these questions, instead, we should critically debate issues and articulate the types of values we are seeking to protect.

Common law engagement with the technology of money

Money is of course a technology and as it has changed across time, it has raised new challenges for law. Money dates from antiquity and is central to organised living. In agricultural communities of premodern time, societies could rely on barter as the main method of exchange. However, as societies and trade became more complex, coins and then banks and bank notes were developed to provide a standard item that could be used as a medium of exchange or value.[35] Modern societies could not exist without monetary systems. Money has many different functions, including a unit of account, measure of value, medium of exchange, means of payment, store of value, standard for deferred payment, liquid asset, framework of the market allocative system (prices), causative factor in economy and controller of economy.[36]

Historically, law has responded to money in a way that was inconsistent with the way that we use and regard money. Although we tend to use money as a fungible that is where one ten dollar note is as good as another, the law does not. The law regards every note and coin as special—as a coin collector would.[37] This regard of money has implications for the external elements of the offence. Larceny historically protected only those who gave away mere physical possession rather than ownership. For example, if I was selling a car and a person wanted to test drive it but actually planned to steal it, this would be larceny by trick. But if someone asks to borrow money and promises to return it, then keeps it instead, this is not larceny by trick because the victim is not expecting those specific notes and coins back—they have handed over property in the specific notes and coins and hence are not protected by common law. This disjunction in outcomes, where the dishonesty of the accused is the same but the property taken was different, resulted in a series of law reforms creating offences such as obtaining by false pretenses and then fraud offences (considered below).

The common law concept of money also had an impact on the *mens rea* for larceny. If a person borrowed a chair, used it, and brought it back—this would not be theft, as the person did not have the intention to permanently deprive.[38] But, if we 'borrow' one hundred dollars from a friend, with the idea that we will return the equivalent value, the law regards this as an intention to permanently deprive the owner—and thus constitutes *mens rea* for theft. There is no difference in the moral blameworthiness of the person intending to return a specific item and the person intending to return the same value in money, particularly because the harm (or lack thereof) to the owner is the same in both examples. The case of *Feely* provides an example of a way in which the courts have sought to handle the question of whether 'borrowing' money with the intention of repaying the equivalent value is sufficiently culpable to justify blameworthiness.[39] In *Feely*, an employer had told all managers that the practice of borrowing money from the tills was to stop. A month after this, the accused took thirty pounds from the office safe to give to his father. He told police that he intended to pay the money back and that the firm owed him seventy pounds in wages. At first instance he was found guilty of larceny, but on appeal, Lawton LJ held that it was a question for the jury whether the defendant's behavior was 'dishonest'. Feely was acquitted as he honestly intended to repay the money and thus lacked the moral obloquy necessary for a criminal conviction.[40] An alternative approach is to regard a person who 'borrows' an item without consent, with the intention of using it as their own, but then returning it, as committing theft or larceny. This would mean that 'borrowing' money without consent but with every intention of returning its value, or purchasing clothes from a shop, wearing the clothes to a party, then returning them for a refund, would be regarded as theft or larceny. Arguably, the example of 'borrowing' clothes from a shop with the intention of returning it would be sufficiently dishonest according to community standards to justify criminalization.

More recently, the law has been grappling with the issue of electronic money—cyber theft. Cyber theft is broadly defined as using an on-line computer system to steal someone else's property or to interfere with someone else's use and enjoyment of the property.[41] Cyber theft provides an example of an intersection of technological developments enabling new methods of stealing new things.

New methods of dishonestly taking

There has been uncertainty in law about how to approach the use of computerised banking systems to take money that the user is not entitled to take. In *Kennison v Daire*,[42] the High Court of Australia considered a case where Kennison had closed his account with the Savings Bank of South Australia but had kept his ATM card. Before the card was disabled, he used it to withdraw $200 from an ATM, knowing that he had no right to the money. The ATM paid him the money because it was off-line and was programmed to dispense $200 cash to anyone correctly entering the card's PIN. Kennison was charged with larceny. He argued that the bank had consented to the passing of property in the money through its programming of the computer. The High Court held that if a bank teller had paid the money to Kennison personally, there would be no larceny because the teller as agent of the bank had consented to the passing of property. However, the High Court held that machines cannot give the bank's consent because 'there is no principle of law that requires it to be treated as though it were a person with authority to decide and consent.' For the High Court, agency to consent requires a human. Accordingly, Kennison was found guilty of larceny.

There are several problems with the 665-word judgment by the High Court. First, there is a failure by the High Court to deal with developments in machines and computers across time. The High Court referred to *R v Hands*,[43] a late nineteenth century case in which the defendant obtained cigarettes from a dispensing machine by using a brass disc instead of the required penny. The court held that this amounted to larceny. The High Court accordingly failed to take into account the difference between a simple mechanical machine and the greater complexity of an automated computer system that has decision-making capabilities. Second, the failure to recognise the complexity of the automated computing system meant that the High Court did not take into account the very small functional difference between ATMs and human tellers. As the criminal law theorist Alex Steel points out, 'both dispense cash based on an authorization given by the bank's computer system'.[44] Banks operate on the basis that there is no difference between a withdrawal via teller or ATM, yet the law regards them as fundamentally different. Third, the High Court leaves the customer in some uncertainty as to whether or not a transaction with an ATM is valid. In contrast, Steel argues that a central value should be the protection of innocent customers from mistakes that it is in the bank's power to control. The failure to recognise that ATMs can consent fails to recognise the imbalance of power between banks and customers, and places the onus on customers to ensure that transactions are correct, rather than banks. The risk falls on the customer rather than the bank, particularly as the bank can pursue the customer civilly. In reality, customers are not able to look behind the apparent consent that the ATM represents, this is particularly the case now that customers can withdraw from the ATMs of other banks and independent ATM operators. Instead of focusing on whether or not the ATM could consent, the law would have been better off focusing on whether there was intentional dishonesty in inducing the consent by the customer, and if there was dishonesty, regarding this as a case of fraud rather than larceny. However, the idea of regarding dishonestly obtaining financial benefits from a computer as fraud raises its own set of difficulties. Can machines receive a representation and thus be deceived? What would amount to a representation? Is it sufficient to wave a card at a machine to amount to a false representation that one is entitled to use that card to pay for goods and services? Steel recommends that the solution to these questions requires the law to engage with interactions with automated systems using specific legal definitions. In these cases, criminality is based on the dishonest attempt to take advantage of the limitations of the programming of the automated system or circumstances that change the nature of the interaction.[45]

Another example of cyber theft is the stealing of money from bank accounts electronically. This form of cyber theft is similar to the old offence of robbery. Both cases involve an owner being unlawfully deprived of property, but they are different operationally. Banks have long implemented prevention procedures against robbery (many of which include technological solutions such as bullet proof glass, surveillance, safes), which have led the offence to have low profitability and a high chance of being caught. In comparison, cyber theft is highly profitable with a low risk of being caught and punished, particularly for offshore offenders. Unlike bank robbery, in cyber theft, the crime scene is scattered across continents. An essential element of robbery is the use or threat of violence. In comparison, cyber theft does not require physical presence or the threat of violence. Brenner has argued that because cyber thieves are unlikely to be caught, the penalties for the offence should be increased, so that the calculation of risk and reward is recalibrated.[46] This would underscore the seriousness of the theft, the potential for cyber theft to erode the security and stability of the financial system. On the other hand, theorists such as Green argue that cyber theft is not more blameworthy than robbery, particularly due to the absence of violence, and different penalties should not be given just because offenders are more difficult to catch. The fundamental essence of criminal law is establishing that an accused is sufficiently culpable to justify the imposition of criminal sanctions.[47] Should criminals' use of cyberspace be treated as an exacerbative harm that is incorporated into definitions of theft?[48]

The advent of fake news provides another example of a new method to profit from dishonesty. Publishers use a two-sided market to profit from fake news on-line. Readers are deceived so that publishers can profit from advertisers. The difficulty is that those who are deceived, the readers, have not lost any money as they have not paid to read the stories. Meanwhile, the advertisers are not defrauded because they are not subject to the deception, they pay the publisher for exposure to the readers and the publisher delivers that exposure (through fake news or 'click bait'). Neither the readers nor the advertisers have been deprived by dishonest means, that is, they have not been defrauded. Although the publisher has not defrauded either the readers or advertisers, it has profited by publishing false statements of fact online. This has led Robert Size to argue that publishing fake news should be regarded as a new way to obtain money by false pretenses, such that it should be regarded as wire fraud.[49] That is, publishing fake news is a scheme or artifice for obtaining money by means of false or fraudulent pretenses or representations. Publishing fraudulent news in a physical newspaper for profit would be regarded as fraud, because it relies on a traditional buyer and seller business model. The seller of a hard copy of fake news is selling it to readers directly. In contrast, fake news online is provided to readers free of charge, revenue is generated by advertising. This is a two-sided market model that is common to the internet. For example, we use Google's search engine for 'free', but advertisers pay to advertise to the users. Google must attract both users and advertisers to profit in the two-sided market for online search. The two-sided market model has meant that most legal analyses do not mention fraud as a possibility for fake news. This kind of fake news can be regarded as a form of social harm, with the sheer prevalence of this kind of deception making it hard for even astute readers to determine whether or not a story is real.[50]

While it may seem extreme to prosecute fake news as fraud, the law has a history of incrementally expanding the types of property that are protected by larceny and theft. The issue of payment of recording artists provides an example of a campaign to change social conceptions of value, but also shows the need for changes taking into account corporate interests. As a consequence of prolific practices within society of illegal downloading of songs and films, governments passed (fairly draconian) legislation criminalizing digital 'piracy' and created advertising campaigns (which showed before films) with 'You wouldn't steal a car...' to shift social conceptions of illegal downloading as acceptable and to regard it as criminal and immoral. This is in accordance with the expressive conception of criminal law, whereby it symbolically communicates, backed by sanctions of the state, right and wrong.

Spotify was developed partly as a response to digital piracy, because one of its founders, Ek, believed that it had significant downsides, including potential for viruses, lack of user friendliness and because people did not like being pirates.[51] Interestingly, though, this shift did not improve payments to artists by record labels and streaming services like Spotify (founded in 2006). Spotify has deals with major record labels such as Universal Music Group, Sony Music Entertainment and Warner Music Group to access their sound recordings on demand and lower the royalty rate to be paid per stream.[52] The deal makes some sound recordings exclusive to Spotify Premium users, lowers the operating costs of Spotify and encourages users to pay the subscription fee to join Spotify Premium, increasing the profitability of Spotify. The major record labels negotiated this deal—in which they receive lower royalty payments and their artists have less exposure—because they own a significant portion of Spotify's stock and accordingly stand to profit from the agreement. The record labels offer reduced royalty rates to streaming services like Spotify, Rdio and Soundcloud in exchange for equity in the streaming companies. However, the artists who make the music receive reduced benefits because their royalty rates are lower and they are available to a smaller market. The artists entered into contracts with the record companies on the basis of the understanding that the record labels had a common interest in achieving the highest royalty rates, but they have no legal recourse against the record labels for these deals which lower their royalties. Basically, the record labels have profited at the expense of the artists who created the music. The courts have declined to rectify the situation because the artists signed contracts which gave the label the right to use its discretion in determining how to exploit the sound recordings.

In 2014, Taylor Swift drew attention to the issue by withdrawing her music from Spotify because she believed she was not being fairly compensated. She was followed by other artists such as Kanye West, Radiohead and Gwen Stefani. Spotify has a complicated payout system that is not publicly disclosed, but it is estimated that a signed artist earns approximately $0.0011 per stream and for an independent artist with over a million streams the average payout is $0.004891 per stream.[53] When the record companies agree to a lower royalty rates it depresses the royalty rates in the market and forces other artists to agree to lower rates. A narrow response to the recognition that artists are not being recompensed adequately is to argue that the Copyright legal regime needs to be reformed to 'keep up with twenty-first century technology' so that artists get paid when their songs are broadcast (on the radio).[54] But this response does not fix all music licensing problems, particularly as these reforms tend to focus on radio stations, rather than the high percentage of royalties music publishers and record labels take from artists in their contract deals. Lesser proposes an alternative approach which is to draw on equity and to argue that the modern relationship between artists and record labels imposes a limited fiduciary duty upon labels to enable artists to recover a portion of profits from the sale of Spotify stock.[55] A fiduciary duty is a product of public policy aimed to regulate opportunism and abuses of trust or confidence. Lesser argues that the artist-label relationship should be considered a partnership, with the label and the artist both sharing in the profits and losses from their venture, and presently, record labels are being unjustly enriched by accepting equity in exchange for a lower royalty rate without sharing the equity with the artists. An alternative approach would be for the law to recognise that this is a form of dishonestly taking, based on inequalities between record labels and desperate artists. It could also be recognised as a form of fraud – yet remains a legal blindspot. This is analogous to the 'fees for no service' scandal in Australian banks – where banks were charging fees for services that were not provided and that was partly enabled by technology—i.e. through automatic payments. It became difficult to track whether services were provided or not/client alive or dead—but ultimately businesses profited at the expense of individuals.

These new forms of profiting through dishonest methods by companies provide examples of the criminal legal system's long term failure to adequately engage with corporate criminality.[56] It is as though the law has a blind spot in relation to dishonesty by corporations. Underpayment of wages, charging where no services have been provided, profiting by undercutting artists and selling advertising in fake news are examples of dishonesty where the community as a whole loses out whilst corporations profit. One argument is to return to the original concept of property offences as criminalizing dishonesty and offences that undermine the community as a whole. Interestingly, the common law and legislature have a long history of engaging with the question of the definition of dishonesty. It is clear that the law attempts to grapple with profiting from objective dishonesty—that is dishonest according to the standards of ordinary people—but until recently many jurisdictions included a subjective component that the accused knew that it was dishonest. When unearthing fees for no service by banks recently, the Australian government believed that this subjective requirement may militate against banks being found guilty of dishonesty offences because the practice was so widespread so banks and bankers may not believe it was dishonest. This led to the Australian government to reform the definition of dishonesty for corporate offences.[57] Despite these law reforms there have been no prosecutions of banks for fees for no service.

New things to take dishonestly

Technology has also created new things to take, including the forms of electronic money discussed above. One question is whether it is possible to take another's identity. Identity theft is the taking of another's identity, while identity fraud is what you do with it. The American reality-based documentary television series *Catfish: The TV Show* investigates the use of false identities in online dating. A catfish is someone who creates a fake identity online by using someone else's pictures and/or false biographical information. The term catfish is derived from the title of the 2010 documentary, in which the film-maker Schulman discovers that the woman with whom he has been having an online relationship has lied in describing herself. The internet facilitates the ability to steal someone else's identity, but the idea of taking someone else's identity is not limited to contemporary times. Patricia Highsmith's novel, *The Talented Mr Ripley* (1955), is about Tom Ripley, a charming criminal who murders a rich man and steals his identity. Identity theft is frequently conceptualised as a financial offence, where, like Tom Ripley, a person obtains another person's personal identifying information to enrich themselves. The victim in this case is the person who suffers financial harm. But what of the person who steals an identity so that another person will fall in love with that false identity, or using the false identity to have on-line sex with other people.[58] Who is the victim in these cases, the person who has suffered 'identity theft' or the person who believes in false identity? The difficulty in these cases is that what is taken is intangible. Nonfinancial types of identity theft raise questions about whether or not my identity has a value that can and should qualify for protection under the law of theft. The law has historically recognised false impersonation offences, such as false impersonation of a police officer, but what of someone falsely using my good name? The contemporary assumption in law is that infliction of financial harm is most serious. One possibility is to criminalise defamation, reputational harm, but this fails to address the residual harm when the imposter does not in/advertently use the victim's identity to harm the person.

Theft causes economic harm, while defamation causes reputational harm but doesn't involve theft because there is no infringement of proprietary interests. The issue of impersonation has been recognised in rules created for social media such as Facebook and Twitter: 'You may not impersonate individuals, groups, or organizations in a manner that is intended to or does mislead, confuse, or deceive others.'[59] Until recently, it was very difficult to credibly pose as another person (but not impossible), now that is quite easily done through social media and dating apps. This has led Brenner to ask whether this 'warrants expanding the law of theft to encompass our names.'[60] Green argues that theft should only apply if a thing is commodifiable (capable of being bought and sold) and rivalrous (the consumption of it by one consumer will prevent simultaneous consumption by others). Under Green's formulation, a victim/owner must lose all or substantially all of what the thief gains.[61] But fundamental, too, to law's construction of crime is that it disturbs the community's sense of security. This may be very serious harm—for example, homicide—or lesser harms such as property offences which nonetheless are more common and so may disturb sense of community due to 'higher likelihood that such lesser harms will be inflicted on us by those who manifest disregard other people's ownership'.[62]

The idea of our personal information and identity having value has become increasingly apparent with the way that our information is (mis)used by corporations for the purpose of profit. In 2018, millions of Facebook users' personal data was harvested without consent by Cambridge Analytica, primarily for the purposes of political advertising. The data was collected through an app, "This is your digital life", that consisted of a series of questions to build psychological profiles on users. Cambridge Analytica arranged an informed consent process for research for 270,000 Facebook users who agreed to complete the survey, solely for academic use. However, Facebook allowed the app to harvest personal information not only from survey respondents but also the respondents' friends, with Facebook later confirming it actually had data on potentially over 87 million users. Through this process, Cambridge Analytica acquired data from millions of Facebook users, enabling Cambridge Analytica to create psychographic profiles on the subjects of the data and their locations. Cambridge Analytica sold the data of American voters to political campaigns and provided analytics to the Ted Cruz and Donald Trump campaigns. For a given political campaign, each profile's information suggested what type of advertisement would be most effective to persuade a particular person in a particular location for some political event. Facebook's database of personal information was used to spread tailored, false political advertisements. While political lies are as old as politics itself, the effect is exacerbated and harder to detect when it is informed by and draws upon data-rich processes. Facebook CEO Mark Zuckerberg apologised for the Cambridge Analytica scandal in 2018 stating that it was a 'mistake' and a 'breach of trust' and promised to make changes and reforms to prevent similar breaches. In April, Facebook implemented the EU's General Data Protection Regulation in all areas of operation, not just the EU. Facebook was fined 500,000 pounds for failing to safeguard people's information by the UK's Information Commissioner's Office in July 2018, the maximum fine allowed at the time of the breach. In July 2019, the Federal Trade Commission voted to approve fining Facebook approximately $5 billion to settle the investigation into the data breach.

One issue for legal theorists and legislators is the question of what precisely is the harm of data misuse? One argument is that it is a personal harm because it breaches the privacy interest of individuals. But a difficulty with this argument is the so-called 'privacy paradox'. That is, however much we say we care about privacy, people frequently treat private personal information with indifference, trading it for access to services from navigation and communication services and participation in social networks.[63] Additionally, the privacy paradigm fails to take into account that the harms from data misuse may be greater than the sum of private injuries to the individuals whose information has been taken. This has led Ben-Shahar to argue that rather than (solely) perceiving data breaches as a private harm, it can be argued that they cause direct and concrete harm to the public ecosystem.[64] Ben-Shahar asserts that 'digital information is the fuel of the new economy. But like the old economy's carbon fuel, it also pollutes'. In the Cambridge Analytica scandal, even though some individuals may have been happy to share their personal information, the data was misused to undermine and degrade the American democratic ecosystem (and to provide targeted fake news). Accordingly, the law needs to protect not just individuals from commercial players in the public sphere, but to protect the public sphere from individuals sharing their personal data with commercial players. The problem is thus not commercial interests misusing personal data in the absence of adequate consent by individuals or failing to adequately compensate individuals whose data has been used. Rather, the key problem is the *negative externalities* through the use of personal data. Ben-Shahar argues that this should be conceived as data pollution. These ideas of misuse of data return us to one of the original ideas underlying the criminalisation of larceny, that it disturbed and undermined the community. It also returns to the classic notion of criminal law as focused around public wrongs and harms. Ben-Shahar argues that like industrial pollution, private causes of action are inadequate because of the public nature of harm. It affects the entire environment, not merely the individuals with whom the polluter transacted or whose data it emitted.

Conclusion

This chapter has shown the long-term engagement by both civil and criminal law with technological developments. We have taken up the specific history of dishonesty offences to show that, while these have existed for centuries, there is still dispute about the types of values and/or interests that the law is seeking to protect. This complexity in the history of dishonesty offences is not necessarily problematic, but can be used as a way to engage with new methods or things to take dishonestly and to query whether the blunt instrument of the criminal law is the appropriate domain to protect specific interests. One key argument that has arisen in this chapter is the blind spot that the legal system appears to have in relation to corporate dishonesty. There are a myriad of ways in which corporations have been able to use new technologies to profit dishonestly. Rather than conceptualise these harms on an individualistic basis, we can return to historic conceptions of dishonesty offences as a public wrong that is harmful to the community or society as a whole. When engaging with new technologies, the law can and should balance not only the relationship between state and individuals, but be aware of the increasing power yielded by corporations. The expressive aspect of criminal law can be drawn upon to communicate the type of society that we wish to be—and this is one in which corporate harms are not tolerated as a cost of business but reconceptualised as blameworthy wrongs.

Notes

1. John G Fleming, 'The Pearson Report: Its "Strategy"' (1979) 3 *The Modern Law Review* 1, 250.
2. n2.
3. Phil Harris, *An Introduction to Law* (Cambridge University Press 2007), 241.
4. John G Fleming, *The Law of Torts* (Law Book Corporation 2002), 7.
5. Morton Horwitz, '*The Transformation of American Law' 1780–1860* (Harvard University Press 1979) 22.
6. Mark Lunney and Ken Oliphant. *Tort Law: Text and Materials* (Oxford University Press 2008) 14.
7. Ralph Harrington, 'The Railway Accident: Trains, Trauma and Technological Crises in Nineteenth Century Britain,' *Traumatic Pasts: History, Psychiatry and Trauma in the Modern Age, 1870–1930.* Ed Mark Micale and Paul Lerner (Cambridge University Press 2001), 36.

8. Henry Hollingsworth Smith, 'Concussion of the Spine in its Medico-Legal Aspects,' *Journal of the American Medical Association* 13 (1889), 182–89, 182, cited by Eric Caplan, 'Trains and Trauma in the Gilded Age,' in *Traumatic Pasts*, n7, 68.
9. David Owen, *Philosophical Foundations of Tort Law* (Oxford University Press 1995), 1.
10. Oliver Wendell Holmes, *'The Common Law,'* (Little, Brown and Co 1881), 38.
11. Oliver Wendell Holmes, 'The Path of the Law,' (1897) 10 *Harvard Law Review* 469, 459–60.
12. n11, 474.
13. The law and economics movement started with two articles in 1961: one by Ronald Coase ("The Problem of Social Cost" in *The Journal of Law and Economics*) and the other by Guido Calabresi ("Some Thoughts on Risk Distribution and the Law of Torts" in *Yale Law Journal.*
14. Edward Veitch and David Miers, 'Assault on the Law of Tort,' (1975) 38 *Modern Law Review* 2 139–52, 139.
15. David Owen, *Philosophical Foundations of Tort Law* (Oxford University Press 1995) 223.
16. Peter Cane, *Responsibility in Law and Morality* (Hart Publishing 2002) 24–5.
17. Desmond Manderson, *Proximity, Levinas and the Soul of Law* (McGill-Queen's University Press 2006) 5.
18. Lunney and Oliphant, n6, 23.
19. Danuta Mendelson, *The History of the Tort of Nervous Shock in Australia, 1886–1991: A Study in the Interfaces of Medicine and Law* (Monash University Press 1993), 8.
20. Peter Cane, n16, 12.
21. Peter Cane, n16, (255–56).
22. Jurgen Habermas, *Between Facts and Norms: Contributions to a Discourse Theory of Law and* Democracy (MIT Press 1996) 367.
23. Adam Seligman, *The Idea of Civil Society* (Free Press 1992), x.
24. Desmond Manderson, n17, 31.
25. Desmond Manderson, n17, 5.
26. Rob Davis, 'The Tort Reform Crisis,' (2002) 54 *University of New South Wales Law Journal*, 897.
27. *Prometheus Bound*, Aeschylus, line 506.
28. *Theogyny* 123. van der Laan, above n 7. 22.
29. Stuart Green, *13 Ways to Steal a Bicycle in the Information Age* (Harvard University Press 2012) <https://www-jstor-org.ezproxy.lib.uts.edu.au/stable/j.ctt2jbt7z>.
30. For example, New South Wales retains the offence of larceny: Section 117 *Crimes Act 1900 (NSW)*. The UK replaced larceny with the offence of theft in 1968: section 1 *Theft Act 1968 (UK)*.
31. George Fletcher, *Rethinking Criminal Law* (Little Brown 1978).
32. George Fletcher, *Rethinking Criminal Law* (Little Brown 1978). Chapter One.
33. At common law, intangible property, like electricity, could not be stolen. This position was altered by statute so that electricity can be regarded as property capable of being stolen, deriving from the *Electric Lighting*

Act 1882(UK). The courts did accept that ephemeral or dynamic forms of property could be stolen provided there was a minimal degree of tangibility, thus gas and water could be stolen. E.g. *Ferens v O'Brien* (1883) 11 QBD 21 and *R v White* (1853) Dears 203; 169 ER 696. Under the *Theft Act (UK)* property has the broadest possible meaning: s4(1) Property means money and all other property, real or personal, including things in action and other intangible property.

34. United Kingdom. Victoria.
35. Glyn Davies, *A History of Money* (Third, University of Wales Press 2002).
36. ibid. 27–8.
37. Great Aunty Diddies used to iron $5 notes for us for our birthdays. They were special notes that had to be used to buy something special, which became increasingly challenging with inflation and increasingly dangerous as the notes became plastic.
38. Taking other people's cars without any intention of keeping them is such a specific problem that many jurisdictions have introduced joyriding offences—where it is criminal to drive or simply be a passenger in a car that has been taken without the consent of an owner, even if the offenders have no intention of keeping the car.
39. *Feely v R* [1973] QB 530.
40. Inherent in the question of honesty is the reasonableness of the honest intent. E.g. an intention to pay back the money upon being paid a salary that week would be reasonable, whereas an intention to pay back the money upon winning the lottery would not be. We discuss the question of the legal meaning of dishonesty below in relation to corporate dishonesty.
41. Susan Brenner, 'Bits, Bytes and Bicycles: Theft and Cyber Theft' (2013) 47 *New England Law Review* 817.
42. *Kennison v Daire* (1986) 160 CLR 129.
43. (1887) 16 Cox CC 188.
44. Alex Steel, 'Both Giving and Taking: Should Misuse of ATMs and Electronic Payment Systems Be Theft, Fraud or Neither?' (2011) 35 *Criminal Law Journal* 202. 209.
45. ibid.
46. Brenner (n 41). 847.
47. Stuart Green, 'Introduction: Symposium on Thirteen Ways to Steal a Bicycle' (2013) 47 *New England Law Review* 795.
48. Brenner.
49. Robert Size, 'Publishing Fake News for Profit Should Be Prosecuted as Wire Fraud' (2020) 60 *Santa Clara Law Review* 29.
50. Shashi Jayakumar, 'Elections Integrity in Fake News Era : Who Protects, and How?' <https://dr.ntu.edu.sg/handle/10220/46743> accessed 6 June 2019; Lili Levi, 'Real Fake News and Fake Fake News Essays' (2017) 16 First Amendment Law Review 232; Edson Tandoc, 'The Facts of Fake News: A Research Review' (2019) 13 *Sociology Compass* 1.
51. Bryan Lesser, 'Record Labels Shot the Artist, but They Did Not Share the Equity' (2018) 16 *Georgetown Journal of Law and Public Policy* 289.
52. These three record labels made up 62% of global music revenue in 2016.
53. Lesser (n 51).

54. Amanda Whorton, 'The Complexities of Music Licensing and the Need for a Revised Legal Regime' (2017) 52 *Wake Forest Law Review* 267. 269.
55. Lesser (n 51).
56. Brent Fisse and John Braithwaite, *Corporations, Crime and Accountability* (Cambridge University Press 1993); Celia Wells, *Corporations and Criminal Responsibility* (Oxford University Press 1993).
57. *The Treasury Laws Amendment (Strengthening Corporate and Financial Sector Penalties) Act 2019* (Cth).
58. Brenner (n 41). 839–41.
59. https://help.twitter.com/en/rules-and-policies/twitter-rules
60. Brenner (n 41). 858.
61. Green (n 29).
62. American Model Criminal Code. 250.7cmt.at 44 (Tentative Draft No. 13 1961).
63. Benjamin Wittes and Jodie Liu, *The Privacy Paradox: The Privacy Benefits of Privacy Threats* (Center for Technology Innovation at Brookings 2015).
64. Omri Ben-Shahar, 'Data Pollution' (2019) 11 *Journal of Legal Analysis* 104.

3 Balance of power and pushing back through the rule of law

Introduction

Has the balance of power between individuals and the state, between the state and corporations, and between corporations and individuals shifted through technological development and, if so, does the law adequately protect individuals from this shift in power? How can law do better? A related issue is that inequality within society can be created and exacerbated by technology—and here, law needs to deliver distributive and administrative justice to ensure that the vulnerable are protected. This chapter considers a robust reading of the rule of law as a means to hold tech to account. The rule of law specifies that laws apply to everyone, whether government ministers, parliamentarians or judges—there is no such thing as unlimited or unreviewable official power. An integral component of the rule of law is the protection of citizens from the arbitrary exercise of power by the state. In recognition of the immense power that the state can wield, the legal system has developed constraints on government action. For example, in criminal law, it is not enough for the state to simply assert that someone is guilty of an offence. An accused is presumed innocent unless and until the state proves guilt beyond a reasonable doubt.[1] We consider the significance of this onus on the role of the state in both using and regulating tech. In the context of the increasing power of global corporations, we also consider how the rule of law should apply to corporations that are making enormous profits from uses of technology in areas that affect individuals' privacy, safety and autonomy.

Big Data and the power of government and corporations

Governments and corporations are increasingly drawing on Big Data, analytics and AI for profit, management and governance.[2] In the

1990s, the internet in general and platforms such as Google in particular were tools that people used for work or entertainment. Now, when we use these tools, we are in turn being instrumentalised by these platforms for our data. An asymmetric power dynamic has arisen between those who hold, control or profit from personal data (usually the state or large multinational companies) and those who provide personal data. Shoshana Zuboff created the term "surveillance capitalism" to describe a market-driven process where the commodity for sale is personal data, and the capture and production of this data relies on mass surveillance through the internet.[3] Zuboff argues that surveillance capitalism is governed by a logic of accumulation that is distinct from the market capitalism that preceded it. This new capitalism, driven by data, creates a power that Zuboff coins the "Big Other": mechanisms that effectively alienate people from their own actions, while also modifying and generating as well as predicting that same behavior. In doing so, the power of the "Big Other" goes against democratic norms as well as normative capitalism.[4] Zuboff argues that the significance of data, algorithms and AI signifies a "new form" of capitalism[5] and a "new species of power".[6]

For the purposes of our book, there are limits to Zuboff's analysis. This is because for Zuboff, the way to fight through or against these transformations of capital and power is not through collective legal or political push-back, but through tropes that rely on individualism. Zuboff laments the loss of the individual's solitude and the "unfettered imagination",[7] and seeks an individually "inward", not a collective, solution.[8] Although Zuboff sees tech as threatening liberal democracy, she sees this as manifesting mostly in the threat to "sovereignty over one's own life",[9] rather than in institutions and processes. Rather than the masses overturning the means of production, with or without the help of technology, Zuboff inexplicably—considering unprecedented levels of casualisation, fired by technology—calls for a return to the halcyon days of twentieth century capitalism where, she seems to assume, workers were at least guaranteed a minimum wage.[10] However, Part III of Zuboff's book is very helpful in characterising the form of power produced by new tech—"instrumentarian power"—which Zuboff uses, for example, to capture the advance of surveillance capitalism, from online to offline presence, with its embedding of data capture technology into numerous quotidian devices and environments which pervade our world. Neither does Zuboff account for law's role, even complicity, in the development of new tech and its corporate, financial and market forms. Zuboff notes that surveillance capitalism involves tech actors asserting power *beyond* law, creating lawless spaces.[11] But this ignores law's productive, constructive role in creating the corporate, employment, intellectual property, privacy and other conditions possible for this new capitalism.

Law's complicity with capital and power is nothing new. Historians of law and capitalism have argued that law has not only enabled capitalism, but has also protected capitalism itself, as a system.[12] For example, at the same time that tort law developed to remedy the injuries of industrialisation, law facilitated the expropriation, enclosure and displacement of labour that powered industrialisation. As David Singh Grewel argues:

> [C]apitalism ... is a juridical regime. It is a form of the modern "rule of law." ... The effect of this regime is that emanation of commercial sociability we now call "the economy." It is produced as the outworking of legal rights and duties that offer special protections to asset-holders legitimated through a constitutional order.[13]

The lawlessness trope of law-and-tech has been supported in part by the idea that nobody owns data. But in fact, law is supporting tech companies at every turn.[14] Julie Cohen provides an account of law's role in her book, *Between Truth and Power: The Legal Constructions of Informational Capitalism.* Here, Cohen explains how law has permeated different stages of capitalism. Instead of "surveillance capitalism," Cohen suggest "informational capitalism" as the key concept, which explains how changes to corporations, power and government have been mediated by law: for example, although data may not be formally "property," law has constructed de facto property regimes to capture both data and algorithms.[15] The law of informational capitalism comprises changes across fields from trade, contract, privacy, corporation law and property law.

Data collection is often carried out by companies that provide us with 'free' online services such as social media platforms (Facebook, Instagram, Twitter) and search engines (Google). Google was the first company to consolidate surveillance capitalism. Instead of charging people per use, it depends on the acquisition of user data as raw material for proprietary analysis and algorithm production to in turn sell; in a further loop, Google now targets advertising of its clients to users through a unique auction model with ever more precision and success. As revenues grew, Google became more motivated to increase and improve data collection. The main 'Big Other' actors are Google, Amazon, Facebook and Apple. Together they collect and control massive quantities of data which they turn into products and services. The collection of data is highly valuable. In August 2020, Apple became the first American company to be valued at more than $2 trillion in market capitalisation. These companies retain information of search histories and make this information available to state security and law enforcement agencies. Many of the strongest corporations that exist today in the world are basically cultural packages for sophisticated algorithms of artificial intelligence (AI) and machine learning. Controlling large data, machine learning algorithms and the ownership of the models that originated from them will be the subject of twenty-first-century feuding and fighting between governments, corporations and individuals. Big Data will become a more important strategic resource than gas, oil or financial capital.[16]

Government agencies have always collected information about citizens for the purposes of taxation, delivery of services and repression, but Big Data has radically increased the scale, speed and complexity of data collection. Information was previously recorded using analogue techniques, but data has the potential to quantitatively and qualitatively change the types of information governments have about citizens. Not only are governments and corporations able to collect more data—they are also able to make connections across data sets that previously were not possible, and to do this more quickly than has previously been imagined. This has led to an argument that Big Data gives governments new capabilities in terms of capacity, scope and purpose. Governments have leapt onto innovative technologies in a quest for agility and increased efficiency, but there is cause to question governments' relationship with Big Data. Data collection can, of course, hold benefits for the public. For example, in the case of COVID19, use of data has provided insights into the disease, and has helped with tracking infection rates; data collection can also benefit other government services including education and welfare.[17] The collection and use of Big Data in government operations increases efficiency and saves time and costs; the lure is obvious. But the deployment of Big Data does not only facilitate the execution of existing government activities—it's also growing government power.[18] Big Data has led to a radical shift in the balance of power between state and citizen, but also corporations and citizens.

The lure of Big Data for governance and commercial purposes has been demonstrated particularly in China. The Chinese Social Credit System standardises the assessment of citizens' and businesses' economic and social reputations. It aims for a unified record system for individuals, businesses and governments and uses a reward and punishment mechanism, summarised as a numerical score. Monitoring occurs through an enormous network of security cameras—over 200 million security cameras through the Skynet program—as well as AI algorithms, smart goggles and face recognition technologies. The overall score is affected by misconduct ranging from bad driving, jay-walking, smoking in non-smoking areas, to how many video games a person buys. If the score is too low, a person may have difficulty renting a house, taking out a loan, or even applying for employment.[19] It might be tempting to argue that the Chinese Social Credit System exists because of the exceptionality of the Chinese government but the case of Edward Snowden provides a warning that all governments are interested in the data of their citizens. Snowden was a former system administrator for the USA's Central Intelligence Agency (CIA) and an information analyst for the National Security Agency (NSA). In 2013, he gave details for secret surveillance carried out by a number of western democracies—particularly NSA and the British Government Communications Headquarters. It should be noted that snooping and suspicionless surveillance is nothing new, but previously governments used methods such as opening mail, wire-tapping, listening in on phone calls.[20] In contrast, the internet enables the possibility of omnipresent surveillance on an unprecedented scale because of the quality and quantity of the data available and the enhanced capacity to analyse the data.

Compounding this issue is the wider geopolitical context and the ways in which national governments are now jostling over corporate use of data. For example, most recently, Australia, the US and then the UK announced that they would not support Huawei, a government-backed Chinese company, from running 5G. Increasingly, the "home states" of these corporations are relevant to the international geopolitics: 'who will own machines and algorithms that will replace half of the world's workforce within a few years? Who will control them and profit from them?'[21]

Until recently, the inherent limitations of analogue data collection meant that the exercise of government power was (arguably) controlled through processes of administrative law and privacy constraints. But Big Data exponentially expands the possibilities and scope of government, extending them beyond existing frameworks of the law. How should these frameworks change in light of Big Data? One view is that Big Data involves crude exercises of government power—even unwarranted exercises of government power—rather than power exercised for the public good. Galloway argued that the way we understand information-gathering and process as a function of government is "fundamentally altered"; following the changed nature of data-government-corporations, we need an adjustment in the use of government power.[22]

Big Data and inequality

All internet users leave digital footprints, personal data and data about their online behavior, which is collected by corporations and governments to identify, sort and classify users. Information can be duplicated and traded, and algorithms can learn from stored data. Companies collect and evaluate huge amounts of data to produce and sell detailed user profiles. This has the potential to create an uneven distribution of data and knowledge, as only those who can afford expensive access to commercially restricted datasets and the experts who analyse these can undertake research. Governments and large corporations are able to guide and capitalise on technological developments in ways that individual citizens and smaller organisations cannot and as such, technological developments are fundamentally shifting the balance of power.

Early literature noted inequalities between those who had access to digital technology and those who did not. People may be excluded based on being unable to effectively use or capitalise on digital technologies, a form of digital exclusion. For example, people who experience homelessness may be digitally connected but rely heavily on their mobile phones to access these connections, resulting in higher rates of debt and credit shortage than the general population. Inequalities in internet access became particularly apparent in the wake of the COVID19 pandemic with many people in lock-down, experiences of the capacity to work (itself a product of the type of employment, if any) and school from home were influenced by internet speeds and these tended to vary with those in regional areas having slower internet speeds. There needs to be greater recognition of digital inequalities and patterns of exclusion.

It is often the most vulnerable who are at the forefront of digital governance, because they are the groups that engage most with government services. In *Automating Inequality,* Virginia Eubanks investigates the impacts of data mining, policy algorithms, and predictive risk models on poor and working-class people in America.[23] Poor and working-class people have long been subjected to invasive state surveillance and technologies, "welfare polices" and punitive policies that have included the compulsory collection of their data. Throughout the nineteenth century, poor people in the US and UK were quarantined in county poorhouses, with local counties required to take records for the indigent people for whom they were responsible. During the twentieth century, poor people were investigated by caseworkers and social workers, and subjected to extensive documentation and monitoring by the welfare state. In the twenty-first century, poor people are subjected to what Eubanks terms a "digital poorhouse",[24] based on networks of algorithms, databases, and risk models. These systems can pre-emptively decide who gets money and food, housing and medical care. Eubanks claims "Big Brother is not watching you, he's watching *us*".[25] People are targeted not as individuals but as members of marginalised groups: as non-whites, as LGBTQI, and/or because they are poor. These groups are subjected to higher levels of data collection when they access public benefits, when they walk through highly policed neighborhoods, enter the health-care system, or cross national borders. That data acts to reinforce their marginality when it is used to target them for extra scrutiny. Those groups seen as undeserving are singled out for punitive public policy and more intense surveillance, and the cycle begins again. It is a kind of collective red-flagging, a feedback loop of injustice.

As legal theorists and lawyers, we need to look at the bigger picture of how this regulation of marginalised groups plays into structural economics. The Global Financial Crisis, and the economic consequences of COVID19 have led to great financial inequality and vast insecurity—all accompanied by the rapid rise in the use of data technology used to track and manage vulnerable populations. A recent example is the case of "Robodebt" in Australia.

Robodebt

Australia's creation and use of the Online Compliance Intervention System (OCI), colloquially known as Robodebt, provides a case study of the potential for the use of automation and/or algorithmic processing to create layers of exclusion to those who are already marginalised and vulnerable, and demonstrates the increasingly asymmetrical relationship between government and individuals. In Australia, Centrelink is responsible for social security payments. The use of data matching in welfare fraud detection is not new, but Robodebt aimed to radically reduce human intervention with Centrelink. Prior to the automated system, a Centrelink Officer would carry out an investigation into any purported inconsistency between a person's reported income and actual income, before sending a letter to the person asking them to provide further details of their earnings in the previous six months.

In mid-2016, the government constructed and applied analytics to uncover overpayments or fraud in relation to welfare payments. The rhetoric of the government in creating Robodebt was to catch 'welfare cheats' through improved efficiency. Prior to Robodebt, there was an average of 20,000 interventions a year. With the introduction of Robodebt, this increased to 20,000 interventions a week, with the capacity to generate up to 783,000 annually.[26] There were calls to stop Robodebt as early as December 2016, due to the number of errors and within and by the system. In May 2020, the Commonwealth government accepted that many debts raised under Robodebt were unlawful and will now refund 470,000 unlawful debts.

The algorithm had major flaws, with the government recognizing (and accepting) a 20% error rate. The algorithm matched Centrelink data with the Australian Taxation Office (ATO) to uncover inconsistencies between the income a person declared to Centrelink and that which they declared to the ATO. A fundamental problem with the algorithm was that Centrelink payments were based on fortnightly earnings, while the ATO recorded annual reported income. This means the algorithm ignored what the person actually earned at the relevant time. A person could work for half a year (and not be entitled to payments during that time), but be entitled to benefits for the other half of that year when they were not working or earning sufficient money. However, the algorithm would then average annual earnings and decide that a person was not entitled to benefits, even though legally they were. Accordingly, Robodebts were unlawful because the amount of social security payments a person is entitled to is based on how much they actually earned when receiving payments, not the averaged ATO amounts. An additional flaw of the algorithm was its inability to account for differences in business names. In many cases, Centrelink and the ATO had listed employer organisations differently (publicly recognised versus formally registered). Although a citizen had accurately reported their income, Robodebt concluded that the individual had double the income that they declared. A more sophisticated algorithm could have been designed to detect that the two different names were in fact the same organization—however, the main problem was not the design of the program, but how the government handled the consequences of its errors. It wasn't only the tech, but the combination of tech and government power that produced the problem of Robotech.

The system, without any human review, automatically generated letters demanding social security recipients themselves prove the calculation was incorrect. Clients were required to prove their Centrelink entitlement by producing pay slips for the period of employment. The calculations could go back as far as 9 years, making it almost impossible to prove the calculation was wrong, particularly as Centrelink advises clients to only keep payslips for the previous 6 months. An additional difficulty was that many did not receive the debt notices because they no longer lived at the address to which the letter was posted. If clients failed to respond or were unable to prove their entitlement with 21 days, Centrelink assumed the debt was correct, and automatically garnished payment from future social security payments or automatically sent and sold the debt to private debt collectors.[27]

Centrelink required repayment within a time limit that expired before the client could offer an explanation. There were not enough staff available for people to provide information or even to ask for assistance. Many calls to Centrelink were simply not answered and Centrelink officers were told not to intervene to halt Robodebt procedures even when human officers suspected the algorithm had incorrectly raised a debt. They were also told not to engage with clients who wished to challenge a debt but rather to direct clients to the online system, even when the online system wasn't working.[28] Clients were required to lodge challenges online, even though the newly constructed webpage was reported to be inoperable at times and very difficult to follow even when it was operable. Those who had not responded within the 21-days' time limit or failed to provide a 'reasonable excuse' via the highly constrained online reporting tool were subject to a 10% debt collection surcharge.

There were many problems associated with Robodebt, and it disproportionately affected the most vulnerable. The algorithm encoded a fundamental lack of procedural fairness in its various procedures. It did not just reproduce less automated procedures overseen by Centrelink staff for determining overpayments—it fundamentally recast legal principles and massively increased the number of people affected. It reversed the burden of proof which should have been on the government to prove the debt; instead, it placed the onus to disprove the debt on the client. The law requires that in order to claim a debt, a debt must be proven to be owed. That is, Centrelink should have proven that a person owed a debt. However, Robodebt shifted the responsibility away from Centrelink to take full responsibility for calculating verifiable debts based on actual fortnightly payments rather than based on an assumed average. The onus of proof placed an immediate financial burden on the client even though there was governmental and community recognition that the algorithm tended to inflate debts. In some cases, it wasn't logistically possible for people to protest those debts within the time limits that had been arbitrarily set by the government—not by the law.

Machine learning algorithms and other digital applications can improve the accuracy and efficiency of decision making but who is responsible when these algorithms fail?[29] There need to be processes in place so that the combination of government power and technology don't replace the rule of law. In *Bureaucratic justice: Managing social security disability claims*, Mashaw outlines the objectives of sound administration, including accuracy, efficiency and dignity.[30] Machine learning initiatives contravene dignity and fairness principles if citizens are disadvantaged by presumed digital literacy and access, lack of understanding of the issue (e.g. the belief that the algorithm is correct and not open to challenge), overcome by feelings of fear and guilt, or impacted by effects that cause a disproportionate impact on members of particular people or groups. Government uses of data must act in accordance with the rule of law, best practice principles and within the principles of ethical consideration of fairness and distributional equity.[31] There were multiple failures associated with Robodebt. There was a failure to ensure that the calculations of debt were accurate in the first place, the government accepted a large number of false positives. There were a large number of people targeted by mismatched data, there were few systems in place to enable people to challenge incorrect debts and they were unable to provide evidence. There was an absence of procedural fairness in carrying out their legislated responsibility. Governments should be assiduous in conducting litigation in ways that avoid oppressive to citizens or are other than consistent with principles of fair play—the Crown is required to act as a model litigant.[32] Robodebt undermined the foundations of the rule of law and interfered with the purpose of government to serve its citizens. In fact, we can think of it encoding the disruption of administrative principles.

Algorithms are both technical and social, and we need to understand their social significance and operation. Technical aspects such as efficiency and accuracy often come to mind when thinking about technology but we also need to think about the context of algorithms, the purposes for the algorithm, the reasons it is created and what it is seeking to achieve beyond technical performance. These are related but distinct questions. In the case of Robodebt, the wider public purpose was portrayed as the government protecting itself (and us) against welfare fraud. Social security payments in the form of the job seeker allowance make up only about 6% of total government expenditure on welfare and social security—so it was in fact bizarre that the government was so energetically pursuing 'debts' in this area. Rather, we should examine this purported purpose skeptically, as a political exercise that was consistent with the language of neoliberalism—the pursuit of individual responsibility and efficiency—while it disguised the bigger issues such its demonstration of a punitive exercise of State power. In this context, data harvesting and analytics and the use of surveillance simply continue disciplinary systems which disproportionately affect the vulnerable and less privileged who are more dependent upon state services.[33]

In the case of Robodebt, people's administrative rights of natural justice and due process were not protected. Initially, the federal government settled legal actions that challenged the validity of the program out of court. In November 2019, the Federal Court declared that the Robodebt of Amato was 'not validly made' because it relied on income averaging.[34] This led to the federal government abandoning sole reliance on income averaging to calculate debts. Finally, in May 2020, after many more cases of incorrect debts surfaced, the government announced that it would repay more than $721 million worth of Robodebt. In addition, a class action seeking interest and damages on behalf of 600,000 Australians is scheduled for September 2020. Questions continue to be asked about how long the government had been aware that its Robodebt scheme was incorrect, illegal and unenforceable. The Administrative Appeals Tribunal had held as early as April 2017 that it failed simple mathematics. Technology provided a veneer of efficiency and neutrality to a scheme that was fundamentally flawed and has ultimately led to more rather than less governmental expenditure.

Robodebt: Legal parameters

In his analysis as both an academic and member of the Administrative Appeal Tribunal (AAT), Terry Carney has called for a return to fundamental legal requirements including the rule of law, transparency of justice and the avoidance of undisclosed settlements.[35] Governments should make (or be compelled to make) the legal reasoning behind any new program initiative public—this could be done through the role of the Ombudsman, or through Parliamentary oversight, which could have a role in monitoring compliance with existing laws. As it stands, there is a lack of legal authority in facilitating this. Part of the difficulty is that the proofs and equations generated by AI are beyond human understanding and imagination and so will need translation by people who have expertise in both tech and legal thinking. One aspect that suffers in the use of advanced algorithms is decision transparency.

It leaves applicants not knowing the basis on which a decision was made and having great difficulty in challenging the reasoning for the decision. Koops has argued that the rule of law requires openness or transparency in authorizing the use of as well as in operationalizing technological management. People should know how decisions were made, there should be periodic independent audits to supplement the accountability of decision-making, and there needs to be safeguards entrenched in the architecture of decision-making.[36] The use of algorithms needs to be consistent with rule of law and due process—the ability to observe, understand, participate in and respond to important decisions or actions that implicate them.

As we noted in Chapter 1, Bennett Moses has argued that there are institutions and systems in place to protect against, challenge and monitor technological disruptions. However, governmental use of Big Data highlights that these institutions and procedures can be undermined by a lack of and/or precarious funding. Despite the continued lure of technology, there have been some large governmental fails. For example, the Australian census collection in 2016 was plagued by technological difficulties. The government suggested problems were due to denial of service attacks on the Australian Bureau of Statistics, but in fact problems arose due to IT failure which had been caused by budget cuts, lack of staff, lack of resources and poor planning. Great skill is needed to engage with data collection. Currently this is undermined at a time of budget cuts. Likewise, leaving aside all the problems associated with understaffing at Centrelink, the government used strategies to avoid higher level legal analysis of Robodebt. Centrelink did not mount contradictory argument before the AAT or appeal against its invalidations of debts as this would have led to AAT2 decision which would be made public.[37] Moreover, this approach individualised the problem and required applicants to mount legal appeals as individuals, turning to the underfunded Legal Aid for legal advice and representation.

As the Robodebt example shows, algorithms are frequently wrong. We explore in Chapter 5 the ways in which algorithms can perpetuate and exacerbate existing discrimination and disadvantage. Another example shows how high the stakes can be. The NSA uses SKYNET, a machine learning algorithm to identify terrorists. However, like Robodebt, there is a fundamental flaw in the algorithm again—there are too few known terrorists in the population to ensure quality of results. Also, the algorithm will not catch terrorists who are different from terrorists who were used to train the model. The algorithm generates false negatives and false positives. If google makes a mistake people see an ad for something they don't want to buy. If the government makes a mistake, they kill innocents or generate mass hardship.[38] Whilst a small error rate may be acceptable for corporations, it is unacceptable for governments.

Advanced algorithms of AI are increasingly being used but these AI functions don't necessarily answer all questions. For example, when programming algorithms of autonomous vehicles, we might have the goal of 'minimizing damage' (a utilitarian approach)—but without further parameters, this goal is too general and unclear. What, for example, would be the order of priorities—should the vehicle choose damage to life over property? What about personal injuries versus damage to property that is priceless or worth millions? Or the injury of many versus the death of one? The archetype here is Philippa Foot's trolley question, who should we choose to save and who should we choose to kill?[39] If the car were programmed to favour the safety of others over the driver and passengers, then people are unlikely to buy that car. The problem is that these questions become difficult, if not impossible, to answer in the abstract: 'the current groping in the dark of AI ethics, based on detached thought experiments as well as suggestions of value-based algorithms, are a dead end'.[40] These kinds of questions are moral dilemmas that philosophers love to debate. The problem is that leaving the decision to algorithms cedes life and death moral decisions to robots. We at least need to have an informed debate about acceptable risks and benefits and what is morally appropriate.

But the use of Big Data also fundamentally changes the balance of power—the extent of government (and corporate) power to collect the information and then how it uses it – data linkage, aggregation and mining and AI 'expands the reach of government power in ways not previously available in an analogue understanding of government activity, for purposes and through processes that are currently unknown and possibly unknowable'.[41] Although the optimistic account of technology conceptualises the internet as a decentralised medium that empowers individuals—it seems instead to have strengthened large organisations resulting in a centralisation of power—favouring centralised solutions. Data is regarded as a valuable commodity, and it is shared by governments and corporations. Aggregate data is of value beyond its value to the individual. It can provide information about ourselves but also the population as a whole. The boundaries between government and private sector collaborations are unclear. Data may be collected for one purpose but then used for additional purposes not originally envisioned. This convergence of data provides a massive boost to government power and with it a validation of power exercised by corporations over the citizen. Zoboff's solution is to argue against these transformations of capital and power through tropes that rely on individualism. However, throughout this book we emphasise the significance of collective legal and political push-back.[42] It is not enough to focus on the costs to individuals in terms of privacy and confidentiality, although these issues are important. We need to think about the protection of community as a whole and the kind of society that we wish to live in.

Notes

1. *Woolmington v DPP* [1935] AC 462.
2. Shoshana Zuboff, '*The Age of Surveillance Capitalism: The Fight for a Human Future at the New Frontier of Power*,' (Harvard University Press 2019).
3. Shoshana Zuboff, 'Big Other: Surveillance Capitalism and the Prospects of an Information Civilization' (2015) 30 *Journal of Information Technology* 75.
4. ibid.
5. Zuboff, n 2, 13.
6. Zuboff, n 2, 5.
7. Zuboff, n 2, 267.
8. Zuboff, n 2, 290.
9. Zuboff, n 2, 513.
10. Zuboff, n 2, 499.
11. Zuboff, n 2, 103–4.
12. Steve Tombs and David Whyte, *The Corporate Criminal: Why Corporations Must Be Abolished* (Taylor and Francis 2015).
13. David Singh Grewal, "The Legal Constitution of Capitalism," in *After Piketty* 471 at 475 (Heather Boushey et al. eds. 2017).
14. See Lothar Determann, "No One Owns Data," (2018) 70 *Hastings Law Journal* 1(5)
15. Julie E Cohen, *Between Truth and Power: The Legal Constructions of Informational Capitalism* (2019), Oxford: Oxford University Press.
16. Tomas Hauer, 'Society Caught in a Labyrinth of Algorithms: Disputes, Promises, and Limitations of the New Order of Things' (2019) 56 Society; New York 222.
17. Lyria Bennett Moses and Louis de Koker, 'Open Secrets: Balancing Operational Secrecy and Transparency in the Collection and Use of Data by National Security and Law Enforcement Agencies' (2018) 41 *Melbourne University Law Review* 530.
18. Kate Galloway, 'Big Data: A Case Study of Disruption and Government Power' (2017) 42 *Alternative Law Journal* 89.
19. https://www.abc.net.au/news/2020-01-02/china-social-credit-system-operational-by-2020/11764740. Yongxi Chen and Anne Cheung, 'The Transparent Self under Big Data Profiling: Privacy and Chinese Legislation on the Social Credit System' (2017) 12 *The Journal of Comparative Law* 356.

20. Glenn Greenwald, *No Place to Hide* (Penguin 2014).
21. Hauer (n 16).
22. Galloway (n 18).
23. Virginia Eubanks, *Automating Inequality*: How High-Tech Tools Profile, Police and Punish the Poor (MacMillan 2018)
24. Eubanks, n 23, 6.
25. Eubanks, n 23, 8.
26. Richard Glenn, 'Centrelink's Automated Debt Raising and Recovery System' (Commonwealth Ombudsman 2017) 2/2017.
27. Paul Henman, '*The computer says 'DEBT': Towards a critical sociology of algorithms and algorithmic governance.* Data for Policy 2017: Government by Algorithm?, London, United Kingdom, 6-7 September, 2017.https://doi.org/10.5281/zenodo.884116
28. ibid.
29. Ellen Broad, 'Who Gets Held Accountable When a Facial Recognition Algorithm Fails?' (2018) 34(4) IQ*: The RIM Quarterly* 18.
30. Jerry Mashaw, *Bureaucratic Justice: Managing Social Security Disability Claims* (Yale University Press 1983).
31. Terry Carney, 'The New Digital Future for Welfare: Debts Without Legal Proofs or Moral Authority?' (2018) 1 *UNSW Law Journal Forum* 16.
32. *Shord v Commissioner of Taxation* [2017] FCAFC 167 [167–74].
33. Virginia Eubanks, *Automating Inequality: How High-Tech Tools Profile, Police and Punish the Poor* (St Martin's Press 2018).
34. *Amato v The Commonwealth of Australia* Federal Court of Australia, General Division, Consent Orders of Justice Davies, 27 November 2019, File No. VID611/2019 (Consent Orders).
35. Carney (n 31).
36. Bert-Jaap Koops, 'On Decision Transparency, or How to Enhance Data Protection after the Computational Turn' in Mireille Hildebrandt and Katja De Vries (eds) (Routledge 2013). 212–3.
37. https://auspublaw.org/2018/04/robo-debt-illegality/ Terry Carney, 'Robo-Debt illegality: A failure of rule of law protections' AUSPUBLAW 30 April 2018.
38. Christian Grothoff and Jens Porup, 'The NSA's SKYNET Program May Be Killing Thousands of Innocent People' [2016] *Ars Technica* 10.
39. Philippa Foot, *Virtues and Vices and Other Essays in Moral Philosophy* (Oxford University Press 1978).
40. Hauer (n 16).
41. Galloway (n 18).
42. Zuboff (n 3). 85–6.

4 Big tech, the corporate form, and self-regulation

Government and the public are increasingly calling not only for the regulation, but the "breaking up" of Big Tech, a call most recently made by American Senator Elizabeth Warren.[1] Public suspicion of organisations including Amazon, Google and Facebook has in many cases rightly arisen as a result of the harms these companies have caused to consumers and more fundamentally to the values of democracy, e-safety and privacy. As individuals, we have become dependent on them for our working and social lives and yet we have little to no capacity to independently remedy issues against them when they arise. Normally, we would look to state law to protect us as consumers, workers and from attacks on our reputation and person, but national governments are showing themselves unwilling or unable to regulate these global corporations.[2] As we argued in Chapter Three, artificial intelligence and automated decision-making tools are increasing in power and tech companies have huge troves of private data that can be relied upon by the state and/or sold on. These companies are at the forefront of technological innovation, and may be caught up with the factual question of what *can* be done, as opposed to the normative question, of whether it *should* be done. The products and processes of tech companies are so new and fluid, and the threats that they pose to society are so invisible, that the legal system is struggling to impose sufficient values and restrictions.[3] Further, as we investigated in Chapter 2, there is an historical complicity between state law, technology and the corporate form.

Accordingly, there is an urgent need for a coherent approach to addressing the ethics, values and moral consequences of responsibility. In this chapter, we look at the relationship of the corporate form to Big Tech. We focus specifically on Google as a case study to analyse the corporate form and investigate whether or not it is capable of working for the public good or if it is doomed by its history and structure to be a/immoral. These questions are extended in Chapter 5, when we look at ethical and legal questions around AI, big data and bias. Of course, Big Tech is not the only story of technology in the twenty-first century—there is also the more utopian idea of tech as disruptive in a pro-democratic way, that tech provides the means of working towards a level playing field *against* more powerful players. Instead of focusing on large multinationals, these disruptive theories tend to be based on small, low-capital innovators that use tech to instigate new forms of relationship and change, including, in the legal sector, NewLaw.[4] However, here we focus on Big Tech because this affects more people—disruption, although sometimes admirably resistant, is still occurring at the margins, while multi-nationals continue to encroach on our daily lives in a monetised way—that "ship has sailed"—and governments around the world are deploying automation tools in making decisions that affect rights and entitlements. The interests affected are very broad, ranging from time spent in detention to the receipt of social security benefits. As we noted in Chapter 1, although technology is lauded as a neutral tool (claims that are challenged by some), this chapter provides an example of the ways in which power and politics are at play in techno-becoming.[5]

Negotiating "Evil": Google, Project Maven and the corporate form

Google has grown exponentially from its origins in a garage to now being the window through which most internet users see the web,[6] surpassing Coco-Cola as the world's most recognised trademark.[7] Google (now part of the Alphabet Corporation) was incorporated on September 4, 1998 as a privately held company operating in a California garage. Two Stanford graduates had invented an ingenious way to rank web pages based on how many other pages link to them. Google's initial public offering was on August 19, 2004, at least three years after the motto 'don't be evil' was coined and adopted. They warned prospective shareholders on the day of its initial public offering that Google was willing to 'forgo some short-term gains' in order to do 'good things for the world'.[8] At the same time, in 2005, Eric Schmidt, former Executive Chairman of Google was quoted as saying, '[e]vil is whatever Sergey [Brin] says is evil.'[9] As of May 2017, the motto was superseded by Alphabet with: "Employees of Alphabet and its subsidiaries and controlled affiliates should do the right thing—follow the law, act honourably, and treat each other with respect".[10] While 'don't be evil' can be interpreted broadly to include not only legal, but moral and ethical concerns, the new motto narrows the ethical obligations and requires only that subsidiaries and employees 'follow the law,' aspirations that provide no guarantee that behaviour is moral. The exact status of the mottos are open to debate—they can be regarded as marketing slogans, or as aspirations and values. They have also formed the basis of codes of conduct. The motto was one of the most controversial, even mythical things about Google.[11] The motto has been variously applauded and disparaged as an albatross,[12] and has been used as a standard to judge Google both from within the institution and externally.

Google has been imbricated in a number of ethical challenges since it was formed. Google's claim to be an algorithm-powered and neutral intermediary between the user and the collective mind of the Internet is no longer true. Google has incentive to game its rankings and there is evidence it does so.[13] Google has been criticised for a conflict in interest: most of Google's revenue comes through advertising from websites subject to its ranking.[14] Given the absence of any comparatively influential alternative, Google has been able to inflate its price for advertising—the highest bidder for a given AdWord will win the war of Internet visibility.[15] This manipulation is exacerbated by Google's aggressive approach to vertical integration, which is the acquisition of companies by other companies so that the acquirer may operate at two or more levels in the same chain of distribution. Since its inception, Google has acquired nearly 100 other companies. This means Google outright owns some of the media content it serves up for searchers, rather than simply just indexing and influencing it. Google gives elevated placement to its own links in the natural results page without subjecting it to the search algorithm, which means it can foreclose the competition.[16] These systems and biases have grown so complex that no Google engineer fully understands them,[17] but nevertheless, Google is carrying on highly political work even where the corporation fails to notice the micropolitics involved. Other ostensible violations of the motto have included issues of free speech, tax avoidance, violation of privacy breaches through Street View,[18] misappropriation,[19] misinformation[20] and the mistreatment of sub-contracted service labourers who are invisible within Google's business model.[21] There has also been emerging disquiet about the use of search engines and social networks for the collection of data of private individuals.[22] Sceptics charge that Google in particular has been heavily complicit in turning the internet into a digital panopticon, making privacy obsolete by treating information as a commodity and turning access to knowledge into a profit-making venture.[23]

Since the mass killings in Christchurch, New Zealand, in March 2019, there have also been debates about the moral obligations of companies, including Google, regarding their role in the sharing of violent, extremist videos without oversight or penalty.[24]

We need to look at new ways of regulating these organisations. We also need to consider the corporate structure that underpins these organisations. Criminologists Steve Tombs and David Whyte argue that existing corporate structures *necessarily* generate a-social, irresponsible outcomes and assert that the corporate form itself should be abolished.[25] Corporate theorists such as Kim argue that, unlike human beings, corporations have no inherent right to exist.[26] This raises the proposition that, where it becomes impossible for a particular corporation to exist without doing wrong—or where we become aware that, increasingly, the corporate form *itself* is not serving human interests—the particular corporation, or the corporate form in general, should not be supported by law and society. In this chapter, however, we take a realist approach and accept the corporate form, for the moment, as given. The question we consider is whether the use of a guiding motto such as Google's "don't be evil" might be one way of negotiating the demands of the corporate form and approaching normative questions raised by technological developments. Does Google's self-appointed framework provide something "more" than lagging legal structures and background ethical frameworks, offering a way forward as we try to navigate a more just realm of technology?

The advantages of incorporation and the public benefit

Wealth maximisation is the default norm in public traded companies.[27] Yet Google conspicuously informed investors upfront that Google would sometimes pursue its social mission, even where doing so might not maximise shareholder wealth. According to Friedman, as a matter of law, corporations *can* benefit non-shareholder stakeholders if that is in fact what the shareholders want.[28] However, in a recent annual meeting for Alphabet, Google's parent company, independent shareholder proposals criticising Google for concentrating power in the hands of a few executives and demanding structural change to make the company more accountable to shareholders and broader stakeholders (including Chinese dissidents) were unsuccessful.[29] So, while technically, legally possible, it seems that this is not normatively taken up as frequently as it should be. Google is not alone in its quest to be, or represent itself as, an ethical business, but the track record of progressively-managed corporations retaining their progressivism across generations is abysmal,[30] which means a live question is whether 'virtuous corporate practices [are] compatible with shareholder capitalism?'[31] Google's motto, "don't be evil", reflects a response to social pressure on high tech organisations to define and disseminate their ethical stances as the influences of technology expand.[32]

'Don't be evil' and Project Maven

In April 2018, thousands of Google employees, including senior engineers, signed a letter protesting Google's involvement in a Pentagon programme that proposed to use artificial intelligence to interpret video imagery and could be used improve the targeting of drone strikes ("Project Maven").[33] More than 3,100 employees signed the letter, setting out their concerns with Google creating warfare technology, with the first sentence stating: 'We believe that Google should not be in the business of war.'[34] The letter explicitly protested in terms of the motto, asserting 'Google's unique history, its motto Don't Be Evil and its direct reach into the lives of billions of users set it apart.'[35] to interpret video imagery that could be used to improve the targeting of drone strikes. Despite ongoing ethical concerns about Google, protests against Project Maven were framed unironically in terms of their motto "don't be evil".

Google has stated that the motto of "don't be evil" partially refers to its commitment to an open workplace of creativity and accountability, requiring a 'culture of tolerance and respect, not a company full of yes men.'[36] It was this openness that enabled employees to challenge Google about Project Maven. But despite claims of an open culture and Google responding to the criticism of Project Maven stating that such exchanges are 'hugely important and beneficial', several Google employees would only talk to journalists on condition of anonymity, stating that they were concerned about retaliation.[37] In addition, while 3,100 employees signed the letter criticising Project Maven, this was only a small proportion of Google's more that 70,000 employees, underlining differences of opinion about the definition of evil and expectations of Google, as well as the fact that not all employees were powerful enough to challenge their employer without fear of retribution. Many of the signatories were senior engineers in the area of artificial intelligence research and thus extremely valuable to the company, so relatively privileged and powerful. In the letter protesting Project Maven, these engineers argued that involvement with Defense would put off the best candidates from working for Google: '[t]his plan will irreparably damage Google's brand and its ability to compete for talent.'[38] Accordingly, the letter was framed in terms that were idealistic/moral but also commercial, appealing to Google's heart and mind within a framework of neoliberal competition.

While Google's response to the employee protests was framed in part to align with the motto of "don't be evil", Google executives also offered various forms of interpretive denial.[39] One form of interpretive denial was offered by Diane Greene, the leader of Google's cloud infrastructure business. She reassured concerned employees that Project Maven was 'specifically scoped to be for non-offensive purposes.'[40] Both Google and the Pentagon said the company's products would not, *in themselves*, create autonomous weapons systems that could fire without a human operator, which was a much-debated possibility in the area of artificial intelligence. Those critical of Project Maven did not accept the 'non-lethal' argument by Greene. The employees' letter in response did not accept this interpretation either:

> Recently, Googlers voiced concerns about Maven internally. Diane Greene responded, assuring them that the technology will not "operate or fly drones" and "will not be used to launch weapons". While this eliminates a narrow set of direct applications, the technology is being built for the military, and once it's delivered it could easily be used to assist in these tasks.[41]
>
> Building this technology to assist the US Government in military surveillance—and potentially lethal outcomes—is not acceptable.[42]

These concerns are legitimate. Defense had, in fact, explicitly asserted that the infrastructure cloud procurement program was in part designed to 'increase lethality and readiness.'[43] This underscores the difficulties of 'separating software, cloud and related services from the actual business of war.'[44]

An additional, even less convincing argument by Google executives was that the technology was not evil because it was intended to save lives by increasing the accuracy of warfare:

> The technology is used to flag images for human review and is intended to save lives and save people from having to do highly tedious work.[45]

This is a consequentialist defence—if the improved analysis of drone video could better identify civilians and thus reduce the accidental killing of innocent people, this can be regarded as a good outcome.

However, the improved analysis could also be used to pick out human targets for strikes. It should be noted the Pentagon already routinely uses analysis of video available on Google in counter-insurgency and counter-terrorism operations, with the artificial intelligence created by Project Maven improving the accuracy of that analysis.[46] The former executive chairman, Eric Schmidt, stated that there was 'a general concern in the tech community of somehow the military-industrial complex using their stuff to kill people incorrectly, if you will.'[47] This expresses an underlying idea that assisting the Department of Defense to kill people "correctly" is not evil. Google already has links with the military: Schmidt is on the Pentagon's Advisory Board, *The Defense Innovation Board* and has defended this position, stating his presence on the Board was to 'at least allow for communications to occur' and that the military would 'use this technology help keep the country safe.'[48]

Whatever the ontological truth about the possibility of good in corporations, the motto 'don't be evil' was of *instrumental* use to employees and formed the basis of their 2018 protest to Google's involvement in a Department of Defence artificial intelligence project.

'Don't be evil' versus the profit imperative: Brenkert's practical approach to ethics

Some ethicists recognise that there are times when moral principles or values clash, and this may require moral compromise.[49] The business ethicist, George Brenkert, considered Google's operation in China, for which it has been heavily criticised, in terms of moral compromise and integrity. Google's ethical question was whether to abandon service to China or to develop a new Google service engine (Google.cn) with servers located in China that would submit to government censorship. Google chose the latter.[50] In response to arguments that this censorship was in breach of human rights, Google's Vice President for Global Communications and Public Affairs, Elliot Shrage, maintained that their decision was compatible with the Google motto of 'Don't be evil' 2006.[51] In a letter justifying Google's continued presence in China, Shrage asserted:

- we believe that knowledge is empowering and that a society with more information is better off than one with less. Providing access to information to people around the world is central to our mission.

- we believe continuing to explore opportunities in markets across the world, including in countries like China, is consistent with Google's mission to organise the world's information and with our commitment to create opportunity for everyone.[52]

This defence of censorship in terms of 'more information' is hypocritical, given Google's commodification of information in search engines through AdWorks. Ultimately, ethical reasoning led Google to withdraw from China, despite the huge profits associated with the service (although Google then contemplated re-entry into the Chinese market with a secret project named "Dragonfly", itself subject to internal dissent by employees due to lack of transparency and the capitulation by Google on human rights, which it in turn abandoned in 2019). Brenkert's position is that we need to be practical and nuanced when considering ethical questions in the context of corporate behaviour. Brenkert considers the situation where realising certain justifiable values may require violation of other justifiable values whose realization can't be made practically compatible with the first set of values.[53] Brenkert argues that if business ethics holds to the importance of human rights and tells Google to go home, then it 'risks being cast as impractical and unrealistic, though idealistic and pure.'[54]

Brenkert attempts to apply ethical principles to corporations while acknowledging that profit is one of their core values. The value of this account is that it requires corporations to make arguments to justify moral compromise. In addition, in this model, the moral compromise is made explicit. The corporation makes the moral compromise because of other responsibilities—particularly profit—but may or may not still guilty of a moral violation, of having done something wrong and against its own values. Most major search engines and social media platforms are owned and controlled by corporate interests that often construe themselves as responsible to some extent to shareholders, employees and other stakeholders and are obligated to follow laws of the states, but are not applauded as being ethical while doing so.[55] Businesses also operate within a market system, so what a business can do is conditioned by the context of what other businesses are doing and its continued existence depends on how well it fares with regard to its competitors. These shifting rules can be seen in claims by Google's engineers that part of the attraction of Google as a workplace is its refusal to do evil.

Brenkert articulates four conditions that should be taken into account when considering moral compromise. We will apply each in turn to Project Maven to consider Google's conundrum of balancing its motto with the quest for profit. We're spending some time with Brenkert because his theory takes ethics seriously, while locating morality within the real world of corporate values. The first condition is the fairness of the particular moral standard being imposed:

> Do the demands of morality place an unfair or unrealistic burden on a moral agent such that the violation or infringement of some moral principle or rule seems warranted even though the principle or rule is otherwise justified?[56]

Brenkert treats morality as a demand and this point raises the question of how much morality can demand of people or corporations. Applied to human beings, this point argues that *any* moral standard that requires individuals to act in ways that would significantly harm or destroy them over a substantial period of time is unacceptable.[57] Using a very particular example of an individual witnessing improper conduct in an organisation, De George argues that morality only demands that a person blows the whistle if he or she has credible evidence of likely harm and has good reason to believe that by going public the threatened harm will be prevented: in the absence of any clear chance of success, whistle blowers are not obligated to blow the whistle.[58] On this argument, there is no absolute duty to a moral standard: there is no need to sacrifice oneself if there is no likelihood positive gain. In alignment with this relative argument would be the view that if Google chooses not to work with Defense, its competitors will—so Google may as well do it. Amazon and Microsoft, for example, proudly promote their links with military and defence agencies. The protest letter directly argues against this point by arguing the exceptionalism of Google:

> By entering into this contract, Google will join the ranks of companies...The argument that other firms, like Microsoft and Amazon, are also participating doesn't make this any less risk for Google. Google's unique history and its motto *Don't Be Evil*, and its direct reach into the lives of billions of users sets it apart.[59]

Brenkert's second condition requires an evaluation of the severity of the violation involved. Harmful consequences are difficult to justify, but aspects that can be taken into account include: What kind of harm does the violation threaten? Does the act add to existing harms? Are innocent bystanders harmed by the violation? The Google employees' protest letter is framed in terms of the harmful consequences of the AI: 'Building this technology to assist the US Government in military surveillance—and potentially lethal outcomes—is not acceptable.'[60] In contrast, the acts of interpretation by Google executives considered in the section above arguably provide support for moral compromise: innocent bystanders would be safer if artificial intelligence could be more accurate, they argued. Only the enemies of the state should fear the drones.

Brenkert's third condition is the implication for one's own integrity. One could argue that moral integrity requires Google to remain absolutely true to its commitments and not compromise at all, and it is this that the employees called for in their protest letter: 'Google's stated values make this clear. *Every one of our users is trusting us. Never jeopardize that. Ever.*'[61] The letter also calls on market values, asking Google executives to protect the brand, a core component of which is the motto "don't be evil". However, in his consideration of integrity, Halfon notes that it is possible for people of integrity to 'reassess their ideals or principles and re-evaluate their commitments. What must be avoided by persons of integrity is abandoning their commitments for arbitrary or capricious reasons.'[62] This could justify Google reconsidering its values in light of the radical changes in the company and the world since it created the mantra "don't be evil". Alternatively, Benjamin has argued that people can maintain their integrity when they accept compromised positions if they protect important values to which they adhere,[63] stating that 'to choose to preserve as best as possible the overall pattern of one's life cannot be regarded as betraying one's integrity. Indeed, in such circumstances, a compromise may provide the best means to preserving it.'[64] On this point, Brenkert has argued that Google could soften the motto, so that it becomes more of an ideal, rather than a rule that can never be violated.[65]

Brenkert's final condition when violating a moral demand is the need to attempt to mitigate the impact or seriousness of the violation. Are there ways in which Google could mitigate and reduce the compromise of the motto by involvement in Project Maven? Arguably, attaching conditions as to how Defense uses the artificial intelligence generated by Google is one way of mitigating harms done, but given that Defense has stated its aim is increased lethality, these conditions are unlikely to be met. The protesters did not accept the idea that Defense will maintain restrictions on use: 'We cannot outsource the moral responsibility of our technologies to third parties.'[66] Instead, the protesters believed the only appropriate response is to cancel the project immediately and 'draft, publicise, and enforce a clear policy stating that neither Google nor its contractors will ever build warfare technology.'[67]

Google, Give up: It is impossible for corporations not to be evil

Brenkert's framework attempts to consider corporations as moral entities within a practical framework, by elevating profit to one of many competing values a corporation may have. Tombs and Whyte would disagree with Brenkert's attempts to portray the corporation as a moral, responsible citizen: their key argument is that corporations are unethically enabled by law and society to make profit.[68] While profit is a value for the corporation, it is not a moral principle. Moreover, as Coffee has argued, corporations are so structured that our standard tools for holding them even to bare legality—such as punishing the legal entity or punishing culpable individuals—suffer from inherent limitations and fail adequately to deter corporate misconduct.[69]

Incorporation is the creation of a legal subject that can be recognised as having a single identity or "personhood" that is distinct from the human persons who make up the corporation. The tension between the motives of profit and social good is inherent in the historical development of the corporation, with Habermas noting that '[a] corporation…may be defined in the light of history as a body created by law for the purpose of attaining public ends through an appeal to private interests.'[70] Historically, from the first application of the corporate form to business enterprise, the reigning norm was that business associates received the privilege of incorporation only on condition that their enterprise would generate public benefits.[71] In the United Kingdom, early corporations were incorporated by special charter following petitions made to Parliament and charters were granted to allow corporations, like the East India Company, to support the state's ambitions for empire.[72] Incorporation was recognised as a substantial privilege bestowed upon a group by the state, giving businesses the significant advantages of operating as a single entity, with jurisdictional authority, centralised management, perpetual succession, asset lock-in, entity shielding and limited liability.[73] If the company incurs losses higher than the value of the sum invested, the owners or shareholders bear no further responsibility for this loss. This in turn means that 'investors will neither be made to pay the full financial losses of the corporation, nor be made to pay for the damage caused by the corporation when social harms are caused.'[74] The advantages that come with incorporation are unavailable to natural persons operating under the general rules of property and contract.

While theorists such as Ciepley recall the historic requirement of public benefit, these theorists also recognise that from the 19th century onwards, the idea that incorporation was justified on the basis of public benefit has been whittled away. This has led the realist economist Milton Friedman to assert that 'there is one and only one social responsibility of business - to use its resources and engage in activities designed to increase its profits.'[75] Friedman clearly articulates the existing paradigm and agenda for corporations, which is the maximisation of profit as both the functional and legal purpose of a corporation.[76] This shift has become axiomatic in the public consciousness, and there is explicit recognition of the primacy of the interests of shareholders and profit maximisation in legal doctrine:

> A business corporation is organised and carried on primarily for the profit of the stakeholders. The powers of the directors are to be employed for that end. The discretion of directors is to be exercised in the choice of means to attain that end, and does not extend to a change in the end itself, to the reduction of profits, or to the nondistribution of profits among stockholders in order to devote them to other purposes.[77]

These effects of the corporation create forms of "structural irresponsibility", where it is often difficult to identify how decisions are made and how well or poorly the different components of production relate.[78] Tombs and Whyte argue that corporate structures necessarily generate a-social, irresponsible outcomes,[79] stating:

> Paradoxically, then, the limited liability corporation – based on the notion of the corporate 'person' – has a *dehumanising* effect. In other words, the primary function of corporate personhood is actually to depersonalise the human consequences of the corporation: to remove its human content. The corporate form removes the necessity for the human consequences of its decisions to be considered at all by its owners.[80]

Interestingly, the Google employees' protest letter is framed in personal terms. The letter is addressed to the CEO ('Dear Sundar') and responses by Google to Project Maven and the motto more generally have been in personal terms. This may underline the exceptionalism of Google, the willingness to name people, rather than to adhere to the myth of a depersonalised corporation. Alternatively, it may reflect the concerns of the recent shareholder meeting about the increasing centralisation of executives at Google. Despite this, Google has shown a willingness to make expedient decisions when the values of profit and "don't be evil" are in conflict.

But are technology companies, specifically, capable of not being evil? There are current views—as well as cultural fears and myths—that there is something inherent to technology that makes tech uniquely likely to exceed human control. For example, one of the founders of DeepMind (which created Evil Genius), Shane Legg, has predicted that the threat of AI running amok is the biggest existential threat to humans this century. Legg asked Google to set up an 'ethics board' to consider the appropriate use of machine learning in its products and to ensure that its new overlord, Google, sticks to its motto.[81] In response, Google set up the Advanced Technology External Advisory Council, but this was almost immediately dissolved due to controversy as to the selection of council members.[82]

Also at play are dystopian myths about technology, portrayed in science fiction films such as *Terminator* and *The Matrix*, which reflect fears that technology dominates or will dominate, all aspects of culture and society.[83] A more nuanced argument can be made that technology companies understand the world and view it through the epistemologies of science and quantifiable data. Journalist and social commentator, Morozov, has drawn upon Habermas to argue that Google has consistently portrayed technocractic consciousness, pointing to the high-profile Google executive proclaiming 'technology is a part of every challenge in the world and a part of every solution.' He also argues that Google is blinded to the non-technological side of human-life.[84] Although its methodology is appropriate for the realms of information theory and computer science, its technological developments have placed Google (and other tech companies) in the moral and social domains of issues such as privacy and information dissemination. As we argued above, there have been a myriad of ethical challenges since its incorporation, which have first surprised Google, and then been handled poorly, if at all. Google has shown a willingness to use its power and clout to expand the reach of technological innovation without considering any negative impacts. The question in relation to developments of AI is not the factual question of whether it *can* be done, but the normative question of whether it *should* be done. This question cannot be answered (only) with quantitative data.

Morozov argues that Google has been blind to the distinction between the factual and normative,[85] stating:

> Time and again, its engineers fail to anticipate the loud public outcry over the privacy flaws in its products, not because they lack the technical knowledge to patch the related problems but because they have a hard time imagining an outside world where Google is seen as just another greedy corporation that might have incentives to behave unethically.[86]

This attempt to reduce the normative to the quantifiable returns us to the contingency of the definition of evil. On this account, even Google's "don't be evil" motto does not protect us from the evil of Google. 'Worst of all,' argues Morozov, 'Google refuses to acknowledge this Kafkaesque dimension of its work, denying the victims of its "algorithmic justice" a way to rectify the situation or even to complain about it.'[87] Google has not acknowledged the political effects of its operations, nor has it accepted that algorithms can be as flawed or biased as the coders writing them, or finding a way to help users report what Google got wrong without have to sue the giant. This is an issue of *information injustice* which cannot be resolved with quantitative data:

> We need to think more carefully about how to regulate to protect values and minimise harm in light of an evolving socio-technical landscape rather than simply asking how technology ought to be regulated'.[88]

Demands that businesses act ethically in international settings have markedly increased, including concerns about sweatshops, environment, transfer pricing and human rights violations.[89] In this way, the *market*, rather than external legal regulation, can possibly govern corporations. However, market regulation has shown itself to be insufficient and is itself dominated by a technocratic, neoliberal consciousness. This is not just an argument about the regulation of technology, but also the corporations that develop and use that technology. There is a large body of academic literature about the failure of the legal system to adequately regulate corporations,[90] and the larger the corporation, the larger the harms it is capable of inflicting and the less likely it is to be held accountable.[91] There is also increasing recognition that the system is failing to adequately regulate technological developments.[92] Combined together, companies such as Google are so new and fluid and the threats that they pose to society are so invisible, that the legal system is struggling to impose sufficient values and restrictions.[93] While it is positive for Google to retain the aspiration of "don't be evil", law needs to impose and protect societal values upon corporations and technological development. Despite its motto, Google has not demonstrated a willingness to impose the limits and values upon themselves.

Conclusion: Some of the ethical challenges to/for Google

Although the motto "don't be evil" can be disparaged as sloganistic and simplistic,[94] there is power in the motto, granted by the framing of protest and response within its terms. The motto reflects a central, founding myth of Google as exceptional, a noble academic exercise that just happens to make money on the side.[95] This arises as part of the anti-corporate myth of the initial creation of Google by two young guys starting an internet tool for the benefit of humans. This origin story no longer holds, as Google has since become hardened into a global corporation. In turn, although the regulation of corporations as a whole may soften this requirement, the effect of the corporate structure is also to make the profit imperative paramount and this imperative is ethics neutral. In addition, companies such as Google reflect and reinforce a technocratic consciousness, elevating quantitative values such as utility and productivity over normative values such as justice and equality. Accordingly, if perfecting drones is the most profitable future direction, then that is where the public company will, or should, go. This is much more the case for a *public* company, versus a *private* company, versus an *unincorporated venture*, because of the profit momentum created by the corporate structure.

Given the arguments by Tombs and Whyte, it may be appropriate for Google to just give up trying not to be evil. It is arguable that the replacement of the motto with "do the right thing" is a form of giving up. Rather than external arguments and evaluations of what it means to be evil, the motto "do the right thing" arguably reduces the standard to simply meaning "do the right thing... for Google". If, as Tombs and Whyte argue, the central purpose of a corporation is profit, then on a teleological argument the best corporation gets the most profit.[96] On this argument, Google should stop the superficial ethical hand-wringing that has largely yielded expedient decisions and simply embrace the pursuit of profit.

However, would we not prefer a corporation that at least aspires not to be evil, rather than a corporation that wholeheartedly pursues and privileges profit above all else (keeping in mind Drucker's argument that the corporation can only flourish long-term if the society in which it is placed likewise flourishes)?[97] The argument regarding Google's aspiration not to be evil versus its realistic ability to act this way is analogous to the argument about law's aspirations to be just versus the realities of the legal system. Jurgen Habermas highlighted this tension between facts and norms, between what the legal system actually achieves and its self-proclaimed ideals.[98] There is often a stark divide between the realists who point to the failings of the system and the idealists who construct abstract models of justice. Habermas contends that even though the legal system does not always achieve ideals such as justice empirically, the ideals are rightly accepted by citizens themselves as engaged participants. This means claims based on these ideals are valid, valuable and powerful, even though the legal system may not (ever) achieve these ideals. This tension between facts and norms is part of common parlance—we make and accept speech claims that promise more than can be possibly delivered in reality. And it is of value that we do so. Even though the legal system may not achieve justice, we still judge the system by the idea of justice and continue to aspire to justice. This suggests that Google can and should retain the motto "don't be evil" as a standard to which the company should aspire, and which would constrain and challenge corporate decision-making: Project Maven was, in face, shut down. So the exceptionalism of tech companies—the unique value of their employees, and the high profile of what they do—paradoxically means that they can potentially lever values outside the profit motive. We should still be calling for greater corporate regulation and oversight of Big Tech—we don't all have the same power vis-à-vis Google as their engineers—but in the meantime, their publicised standards leverage critique, especially from within the organisation.

Notes

1. https://medium.com/@teamwarren/heres-how-we-can-break-up-big-tech-9ad9e0da324c
2. See, for example, Carolyn Abbott, 'Bridging the Gap - Non-State Actors and the Challenges of Regulating New Technology' (2012) 39 *Journal of Law and Society* 329; Lyria Bennett Moses, 'How to Think about Law, Regulation and Technology: Problems with "Technology" as a Regulatory Target' (2013) 5 Law, Innovation and Technology 1.

3. Evgeny Morozov, 'Don't Be Evil' [2011] *The New Republic* <https://hci.stanford.edu/courses/cs047n/readings/morozov-google-evil.pdf>.
4. Joan C. Williams; Aaron Platt; Jessica Lee, "Disruptive Innovation: New Models of Legal Practice", *Hastings Law Journal* 67, no. 1 (December 2015): 1–84; Tania Sourdin; Bin Li; Tony Burke, "Just, Quick and Cheap: Civil Dispute Resolution and Technology," *Macquarie Law Journal* 19 (2019): 17–38
5. Monika Zalnieriute, Lyria Bennett Moses, George Williams. The Rule of Law and Automation of Government Decision-Making. (2019) *The Modern Law Review* 82.
6. Andrew Carroll, 'Don't Be Evil... Unless It Increases Revenue: What the Operation of Credit Rating Agencies Can Teach Us about Google' (2012) 31 *Temple Journal of Science, Technology & Environmental Law* 93.
7. Lastowka, Greg, 'Google's Law' [2008] *Brooklyn Law Review* 1327.
8. Morozov (n 3).
9. ibid.
10. Jo Ann Oravec, 'Mottos and Ethical Statements of Internet-Based Organizations: Implications for Corporate Social Responsibility' (2014) 1 *International Journal of Civic Engagement and Social Change* 37. 222.
11. Morozov (n 3).
12. Hoofnagle, C.J., 'Beyond Google and Evil: How Policy Makers, Journalists and Consumers Should Talk Differently about Google and Privacy' (2009) 14 *First Monday* 4.
13. Carroll (n 6).
14. ibid.
15. ibid. See *Kinderstart.com, LLC*, No. 06-2057.
16. ibid.
17. Morozov (n 3).
18. Oravec (n 10).
19. Brandon J Huffman, 'Developments in Social Media: First Amendment, Privacy, and Misappropriation' (2015) 71 *The Business Lawyer* 305.
20. Gustavo Resende and others, '(Mis)Information Dissemination in WhatsApp: Gathering, Analyzing and Countermeasures', *The World Wide Web Conference* (ACM 2019) <http://doi.acm.org/10.1145/3308558.3313688> accessed 6 June 2019; Siva Vaidhyanathan, *Antisocial Media: How Facebook Disconnects Us and Undermines Democracy* (Oxford University Press 2018); Maxwell Tani, 'Fake News about the Las Vegas Shooting Spread Wildly on Facebook, Google, and Twitter' *Business Insider Australia* (3 October 2017) <https://www.businessinsider.com.au/las-vegas-shooting-mandalay-bay-facebook-twitter-google-2017-10?r=US&IR=T> accessed 6 June 2019.
21. Alison Agnes Veith, '"Don't Be Evil": Google's Labor, Technology, and the Limits of Corporate Good' (Master of Arts in Women and Gender Studies, San Francisco State University 2015).
22. Oravec (n 10).
23. Morozov (n 3).

24. ABC News, "New Zealand PM Jacinda Ardern leans on Facebook to drop Christchurch shooting footage."
25. Steve Tombs and David Whyte, *The Corporate Criminal: Why Corporations Must Be Abolished* (Taylor and Francis 2015).
26. Susanna M Kim, 'Characteristics of Soulless Persons: The Applicability of the Character Evidence Rule to Corporations' (2000) 2000 *Illinois Law Review* 763.
27. Antony Page and Robert A Katz, 'Is Social Enterprise the New Corporate Social Responsibility Berle II: The Second Annual Symposium of the Adolf A. Berle, Jr. Center on Corporations, Law & Society' [2010] *Seattle University Law Review* 1351.
28. Friedman, Milton, 'The Social Responsibility of Business Is to Increase Its Profits' *New York Times Magazine* (13 September 1970). 51.
29. Tiku, "Alphabet Shareholders Demand Accountability." (Wired 2019) <https://www.wired.com/story/googles-troubles-encroach-alphabets-shareholder-meeting/> accessed on 6 January 2021.
30. O'Toole, 'The Enlightened Capitalists - James O'Toole - Hardcover' (*HarperCollins Publishers: World-Leading Book Publisher*) <https://www.harpercollins.com/9780062880246/the-enlightened-capitalists> accessed 23 July 2019.
31. O'Toole, *ibid.*
32. Oravec (n 10).
33. Shane, Scott and Wakabayashi, Daisuke, '"The Business of War": Google Employees Protest Work for the Pentagon' *The New York Times* (4 April 2018) <https://www.nytimes.com/2018/04/04/technology/google-letter-ceo-pentagon-project.html> accessed 8 July 2019.
34. "Letter to Google C.E.O."
35. "Letter to Google C.E.O."
36. Stated by Google's Executive Chairman Eric Schmidt and Chief Economist Hal Varian in *Newsweek in* 2005, quoted in Oravec (n 10). 223.
37. Shane, Scott and Wakabayashi, Daisuke (n 33).
38. "Letter to Google C.E.O." https://static01.nyt.com/files/2018/technology/googleletter.pdf
39. Stanley Cohen, *States of Denial: Knowing about Atrocities and Suffering* (Polity 2001).
40. Clifford, "3,100 Google employees to CEO Sundar Pichai: 'Google should not be in the business of war'."
41. "Letter to Google C.E.O."
42. "Letter to Google C.E.O."
43. Shane, Scott and Wakabayashi, Daisuke (n 33).
44. ibid.
45. ibid.
46. Pellerin, "Project Maven to Deploy Computer Algorithms to War Zone by Year's End."
47. Kosoff, "Google employees 'outraged' their tech is being used to build better killing machines," (Vanity Fair 2018). <https://www.yahoo.com/entertainment/google-employees-outraged-tech-being-233645780.html> accessed on 6 January 2021.
48. Scharre, "Erich Schmidt Keynote Address at the Centre for a New American Security Artificial Intelligence and Global Security Summit."

49. Benjamin, Martin, *Splitting the Difference: Compromise and Integrity in Ethics and Politics* (University of Kansas Press 1990). See also Berlin, *Two Concepts of Liberty,* 169. Other ethicists do not allow for the possibility that values and principles will clash, and thus do not regard moral compromise as ethically permissible.
50. George G Brenkert, 'Google, Human Rights and Moral Compromise' (2009) 85 *Journal of Business Ethics* 453.
51. Quoted by ibid. The issue has recently resurfaced again. See Amnesty International website providing a letter from Kent Walker, Senior Vice President for Global Affairs, 'China: Google response to concerns over project DragonFly and human rights in China.' Defended in human rights terms and consistent with Google's code of conduct. https://www.amnesty.org/download/Documents/ASA1795522018ENGLISH.PDF. Accessed July 8, 2019.
52. Letter from Kent Walker, Senior Vice President for Global Affairs, Google LLC, October 26, 2018 to Amnesty International: https://www.amnesty.org/download/Documents/ASA1795522018ENGLISH.PDF. Accessed July 9, 2019.
53. Brenkert (n 50).
54. ibid. 472.
55. Oravec (n 10).
56. Brenkert (n 50).
57. Brenkert (n 50) 469.
58. De George, R.T., *Competing with Integrity in International Business* (Oxford University Press 1993).
59. "Letter to Google C.E.O."
60. "Letter to Google C.E.O."
61. "Letter to Google C.E.O." (italics included in letter).
62. Halfon, M. S., *Integrity: A Philosophical Inquiry* (Temple University Press 1989).
63. Benjamin, Martin (n 49).
64. ibid.37.
65. Brenkert (n 50). 471.
66. "Letter to Google C.E.O."
67. "Letter to Google C.E.O."
68. Tombs and Whyte (n 25).
69. John Coffee Jr, 'No Soul to Damn: No Body to Kick: An Unscandalised Inquiry into the Problem of Corporate Punishment' (1980) 79 *Michigan Law Review* 386.
70. Adams, H.C., 'Economics and Jurisprudence' (1897) 2 *Economics Studies* 7.
71. David Ciepley, 'Can Corporations Be Held to the Public Interest, or Even to the Law?' (2019) 154 *Journal of Business Ethics* 1003.
72. The East India Company conquered, subjugated and plundered vast tracts of south Asia. Thus although the company was originally incorporated for 'public benefit', the project of colonization has since been challenged: Robins, *The corporation that changed the world how the East India Company shaped the modern multinational.*
73. Ciepley (n 71). 1004.
74. Tombs and Whyte (n 25). 84.

75. Friedman, Milton (n 28).
76. Janine S Hiller, 'The Benefit Corporation and Corporate Social Responsibility' (2013) 118 *Journal of Business Ethics* 287.
77. *Dodge*, 170 NW 668.
78. Pearce, 'Crime and Capitalist Business Organizations | F. Pearce | Request PDF' (*ResearchGate*) <https://www.researchgate.net/publication/313679483_Crime_and_Capitalist_Business_Organizations> accessed 16 July 2019.
79. Tombs and Whyte (n 25).
80. ibid.85.
81. "Don't be evil, genius: Machine Learning"
82. Statt, "Google dissolves AI ethics board just one week after forming it." (The Verge 2019) <https://www.theverge.com/2019/4/4/18296113/google-ai-ethics-board-ends-controversy-kay-coles-james-heritage-foundation> accessed on 6 January 2021.
83. JM van der Laan, *Narratives of Technology* (Palgrave MacMillan 2016). See also Grant, *Technology and the Trajectory of Myth*.
84. Morozov (n 3).
85. ibid.
86. ibid.
87. ibid.
88. Lyria Bennett Moses, 'Agents of Change: How the Law "Copes" with Technological Change' (2011) 20 *Griffith Law Review* 763. 787.
89. Brenkert (n 50).
90. For example, John Braithwaite, 'Taking Responsibility Seriously: Corporate Compliance Systems' in Brent Fisse and P French (eds), *Corrigible corporations and unruly law* (Trinity University Press 1985); Celia Wells, 'Corporate Criminal Liability: A Ten Year Review' (2014) 12 *Criminal Law Review* 849.
91. Scott Veitch, *Law and Irresponsibility: On the Legitimation of Human Suffering* (Routledge 2007).
92. For example, Abbott (n 2); Bennett Moses (n 2).
93. Morozov (n 3).
94. Oravec (n 10).
95. Morozov (n 3).
96. Aristotle, *The Nicomachean Ethics* (J Thomson tr, Penguin 2004). This argument can be applied to the television series *True Blood*. If the vampires are created to drink blood, then rather than feel guilty about it as the character Bill does, the vampire should embrace and enjoy the lifestyle like Eric Northman.
97. Peter Drucker, *Post-Capitalist Society* (HarperBusiness 1993).
98. Jurgen Habermas, *Between Facts and Norms: Contribution to a Discourse Theory of Law and Democracy* (Polity Press, 1996).

5 Critical legal theory and encountering bias in tech

Bias and regulation

In chapter 4, we interrogated the corporate form and investigated whether or not it is capable of working for the public good or if it is doomed by its history and structure to be a/immoral. In this chapter, we extend this question through an analysis of the ethical and legal questions around artificial intelligence (AI), big data and bias. The regulation of AI is an emerging global issue. Questions around regulation generally focus on specific, even technical questions: for example, focusing on managing the risks of autonomous AI, questions of accountability or safety and privacy issues. Outside these specific issues, not much attention has been given to how regulation might respond to the wider, structural issues of bias and inequality in AI and tech.[1]

In 2019, the Global Partnership on AI, modelled on the International Panel on Climate Change, was put forward by six members of the Group of Seven (G7), an international organization comprised of nations with the largest and most advanced economies. The G7's dominant member, the United States, has not yet signed on, arguing that regulation will be too stifling to a developing technology.[2] Some AI corporations are advocating for more government oversight, but most, perhaps not surprisingly, are arguing for self-regulation. The EU (European Union) is notoriously more sympathetic to regulatory solutions to risk than most other jurisdiction. In February 2020, the European Commission released a white paper "On Artificial Intelligence—a European approach to excellence and trust".[3] Bias is not considered in the white paper, but significantly, the white paper *does* consider the more limited concept of "discrimination", asking for:

> requirements to take reasonable measures aimed at ensuring that [the] use of AI systems does not lead to outcomes entailing

> prohibited discrimination. These requirements could entail in particular obligations to use data sets that are sufficiently representative, especially to ensure that all relevant dimensions of gender, ethnicity and other possible grounds of prohibited discrimination are appropriately reflected in those data sets.[4]

This is a worthy goal, but how should it be meaningfully achieved, and what other factors are relevant when considering bias? Further, is regulation the best way through the question of bias in tech?

Critical theory and bias

Critical theory is a tradition of legal jurisprudence that asks how law legitimates power—and how power is legitimated by law. Since the 1990s, theorists such as Kimberlee Crenshaw, Patricia Williams and Richard Delgado have examined the ways structural inequalities—racism, sexism, classism—are perpetrated by liberal law, while often disguised by legal rhetoric and processes. More recently, scholars have examined the material histories of colonialism and slavery to demonstrate law's complicity in oppression—here, see the work of Renisa Mawani, Stephen Best, and Stewart Motha. In this, the final chapter of our book, we look to the tradition of critical legal studies for ways to think through questions of technology and bias. We argue that it is important to not only look at questions of gender and race, but to look at particular material histories through which race and gender are produced—especially the histories of colonisation and slavery—and how these ground the material conditions behind technology, as the best of critical work has taught us to think about law.

AI is capable of wonderful achievements: AI can find missing children by matching data;[5] Google's DeepMind has trained a neural network on a database of scans so that it can diagnose disease by looking at patients' eyes;[6] and AI can tell when people are depressed, even suicidal.[7] AI has been used to detect new types of breast cancer so that oncologists are now able to cure 50% of all cancers. AI—in its ability to access and analyse huge amounts of data—can *potentially* be used to ensure medical research draws on wider demographics than it has in the past. In the research of rare diseases—where there might otherwise be a lack of data—doctors can use AI to tap into global data to understand more about diagnosis and treatments.

On the other hand, facial recognition algorithms studied by Algorithmic Justice League founder Joy Buolamwini found that 80% of input images on which facial recognition algorithms were based were white, and 75% were male. The algorithms were 99% accurate in detecting male faces, but only 65% accurate at detecting the faces of black women.[8] So, focusing on gender only is not likely to solve other intersectionality issues in AI.[9] And while medical research is being accelerated by AI, medical trials have not been representative, with most participants being white, older, wealthy males.[10] Infamously, in 2016, Microsoft launched Tay, an AI chatbot that Microsoft described as an experiment in "conversational understanding".[11] Tay was designed to be a social robot and to learn through engagement—the more people who chatted with Tay, the smarter Tay was meant to become. As soon as Tay was launched, people started tweeting @Tay with racist, misogynist conversation. Within a day, Tay was spouting so much hate-speech, the robot had to be put to sleep. More and more examples of bias and misuse of AI technologies are being revealed—from insurance and health algorithms that exclude vulnerable people from coverage, to police use of racist facial recognition technology—but despite these problems, there is little transparency and no oversight by government or international bodies.

According to Timinit Gebru, society's views of race and gender are "encoded" into AI systems.[12] Gebru writes:

> … the lack of representation among those who have the power to build this technology has resulted in a power imbalance in the world, and in technology whose intended or unintended negative consequences harm those who are not represented in its production.[13]

This includes predictive policing systems, facial recognition tools and automatic gender recognition (AGR) systems. The consequences of this racial bias are profound, with the potential for people to be sent to be searched, arrested or even incarcerated due to mistaken identity.[14] Similarly, Gebru identifies how AGR systems—which are used primarily for targeted advertisement—are often programmed using data that has very few transgender or non-binary individuals and the effects of this range from misgendering an individual to outing them in public.[15] Unlike the optimistic accounts of Haraway and Braidotti that we articulated in Chapter 1, technology thereby reproduces and accentuates structural sexism and racism. Police officers, for example, do not actually need to be 'racist' for facial recognition technology to have a discriminating effect. Similarly, advertisers do not actually need to be 'transphobic' for AGR systems to misgender an individual. Further, the intersectional effects mean that non-white women and

LGBTI people are most affected. Our work in Chapter 5 will be to outline the unique contribution that critical legal theory can contribute to not only critiquing the development of machine-learning, but also in shaping new technology.

Predictive Policing

Predictive Policing refers to any system that uses data to predict *where* a crime may occur (place-based prediction) or *who* will be involved in a crime as either perpetrator or victim (person-based).[16] But far from being scientific or "neutral", these systems produce *dirty data*—data that is racially—biased and that is obtained in turn through *dirty policing*, policing that is biased and that targets minority communities (either in a place-based or person-based way).[17] This relation between data and policing creates a "confirmation feed-back loop" that a number of theorists have commented on: Black communities are over-policed in the first place, which leads to more criminalising data being available, which in turn is used as predictive data to profile Black communities as criminal communities. Which leads to more over-policing. This confirmation feedback loop affects judicial processes from deciding who is prosecuted to corrections processes and sentencing outcomes. The primary (and most obvious) consequence of predictive policing is that police discriminately target certain communities. The reality of this can essentially be summarised by reference to Ruth Wilson Gilmore's definition of racism, which is "the state-sanctioned or extra-legal production and exploitation of group-differentiated vulnerability to violence and even premature death".[18] Predictive technologies are compounding structural violence—especially police brutality and black deaths in custody. State violence through incarceration compounds intersectional violence. Indigenous Australians are the most incarcerated people in the world, and Aboriginal women are the most imprisoned population within that group, despite comprising only 2% of the Australian population.[19] According to research by Sisters Inside, an advocacy group for Australian women in prison, up to 90% of women in prison have been sexually abused and 98% have experienced violence.[20]

AI in medicine

There are several problems with supposedly 'neutral' AI medical technologies. First, the commercial interests of tech companies compound problems with AI: the drive of tech companies to make profit often outweighs the need to provide a representative data. Second, there is no requirement that the data that AI technologies rely upon is ethnically diverse, and there is no systematic method by which companies can obtain representative data.[21] Third, where health care systems have tried to implement data-checking mechanisms (as is the case of the British National Health Service (NHS), where checks are carried out on the "quality of data"), the efficacy of such systems to ensure diversity remains questionable.[22] Fighting algorithmic bias needs to be a priority, as a question of social justice. By 2022, Gartner Inc predicts that 85% of all AI projects will be delivering problematic outcomes because of bias in algorithms, data or due to the teams/institutions managing them.[23] This is a problem for social justice, and it also decreases the usefulness of AI.

Online trolling and bias

If you are engaged with the internet, especially if you're engaged with social media, you will be familiar with online harm—caused by everything from the uploading of footage of the Christchurch terrorist shootings and child pornography to sexual harassment and "trolling". In this section, we focus on the particular online harm of trolling and the unwillingness of the major social media platforms—as well as the state governments that could potentially regulate them—to make any entity accountable for these harms and the uneven ways in which these harms fall. In *Crash Override*, Zoë Quinn explains that part of the problem is definitional, as "trolling" has come to mean anything from teasing people "for a joke" to extreme hate-speech and threats on people's lives.[24] Quinn became famous as the archetypal target of trolling in the Gamergate controversy, which began in 2014. Gamergate began with acts of coercive control, when Quinn's ex-boyfriend committed acts of libel by posting disparaging comments about Quinn on online forums. As a result, Quinn endured a long period of doxing, rape threats and death threats, which in turn became a worldwide, misogynist campaign against Quinn and a number of other women. The campaign occurred across Reddit, IRC and 4chan. Quinn and others targeted in this campaign suffered significant harms, including psychological damage, loss of income and financial cost.[25] Quinn's case is one of the most well-known cases of trolling but this phenomenon has become much more pronounced—and yet law has not yet developed an adequate response. The research of Danielle Keats Citron, a Professor Law at the University of Maryland, suggests that victims are encouraged to simply go offline, ignoring the reality that so much of our work and social opportunities in the 21st century are now online—meaning that going offline means income and opportunity lost.[26]

Part of the problem, Ginger Gorman argues in her book *Troll Hunting,* is the false assumption that the online world is "virtual" and incapable of inflicting real-world harms; Gorman argues that the harm of trolling goes far beyond "hurt feelings"—rather, Gorman claims that what she terms *predator trolls* destroy lives by destroying people's mental health, their livelihoods and even contribute to suicides.[27] Emma Jane argues that one of the particular harms of trolling is "economic vandalism"[28] and that gendered cyberhate in particular is "an insidious new form of workplace harassment".[29] In Amnesty International's major *Toxic Twitter* report, Dunja Mijatović argues that 'the severity, in terms of both sheer amount and content of abuse, including sexist and misogynistic vitriol, is much more extreme for female journalists'.[30] Gorman asked The Australia Institute (TAI) to provide a calculus of harm for trolling and cyber-hate. TAI conducted a nationally representative survey of 1557 people in 2018 for Gorman's book *Troll Hunting* and found 44% of women and 39% of men have experienced one or more forms of online harassment.[31] The Pew Center in the United States found in 2017 that 'around four-in-ten Americans (41%) have been personally subjected to at least one type of online harassment'.[32] Amnesty International's Toxic Twitter report shows the disproportionate ways different people are targeted online. Women of colour, religious or ethnic minority women, lesbian, bisexual, transgender or intersex (LBTI) women, women with disabilities, and non-binary individuals, experience targeted abuse that targets them in compounded ways, as well as in greater numbers than white men.[33]

Essentially, social media platforms have been left to govern themselves. According to Facebook, all posts must comply with their 'Community Standards'.[34] Facebook reported that in the second and third quarter of 2019 that it "took action" against tens of millions of posts, photos and videos for violating its rules that prohibit hate speech, harassment and child exploitation.[35] The "action" cited includes removing or labelling more than 54 million pieces of content that Facebook deemed violent or graphic, 18.5 million items determined to be child nudity or sexual exploitation, 11.4 million posts that broke its rules prohibiting hate speech and 5.7 million uploads that broke bullying and harassment policies.[36] Its level of action increased last year following the Christchurch terrorist shootings on 15 March 2019, when an Australian man attacked two mosques, and which were shared widely over social media, including Facebook.

Beyond these self-governing acts of moderation, Facebook resists the idea that it is responsible for harm caused by online harassment and bullying, presenting itself as an intermediary of content rather than as a publisher. The question of whether social media companies are publishers—and therefore liable for content that is false, misleading or malicious—or not, is a live question.[37] Surprisingly, Amnesty International *opposes* the imposition of legal liability 'for companies who fail to remove abusive content' as it argues it 'sets a dangerous precedent and risks causing more harm instead of addressing the core of the issue' of hate speech and harassment.[38] Emily Bell, director of the Tow Center for Digital Journalism at Columbia University, argues that social media companies are publishers because they 'monetise, host, distribute, produce and even in some cases commission material'. Further, unlike traditional publishers they *amplify* content through technology, promoting the cheapest and "most engaging" content, which might include false, discriminatory and hateful material'.[39] Tech companies have argued that the vast volume of content on their platforms makes it impossible to monitor. Frank Pasquale, author of *The Black Box Society*, argues that these platforms are driven by anger and anxiety, and the companies behind them have incentives to encourage this behaviour.[40]

Should social media companies and other website operators be accountable for content on their sites? Under Section 230 of the US Communications Decency Act of 1996, social media companies are not responsible for anything that third parties publish on their platforms, although it is now marked by the Department of Justice as an area of concern.[41] Section 230 has recently been amended, but only as it applies to sex trafficking.[42] In the early days of the internet, Section 230 was celebrated as promoting democratic free speech and equality for internet citizens. When Section 230 was legislated in the mid-1990s, the intent was to encourage the internet to grow without censorship. Senator Ron Wyden co-authored the legislation and recently stated that the intent behind giving companies the power to self-regulate (moderate) was to provide tech companies with both a "sword" to be able to raise capital and a "shield" to protect them from potential liability posed by third-party users of their sites.[43] But concomitant to those privileges, Wyden says, was the expectation that tech companies properly take care of the safety of users of their platforms:

> [W]hat was clear during the 2016 election and the succeeding events surrounding Facebook, is that technology companies used one part of what we envisioned, the shield, but really sat on

their hands with respect to the sword, and wouldn't police their platforms.[44]

Whitney Phillips argues that trolling is not just about toxic individuals. Rather, it is the effect of the amplification of a problematic culture in general, and profit-seeking social media companies more specifically. For trolls, exploitation is a leisure activity; for media, it's a business strategy. Phillips argues that trolls align with social media.[45] Trolling, widely condemned as obscene and deviant, actually fits comfortably within the contemporary media landscape and is culturally normative, rather than deviant. We can demand change through great regulation of social media companies, but we should also be demanding deeper cultural and social change.

Facebook moderators

The internet is dominated by large corporations who have extreme profit motives for encouraging particular kinds of user interaction—uses that do not align with the values of democracy and equality, and that actively cause more harm to non-white, non-male and non-cis people. Philosophically, then, it is difficult to see why social media companies should be shielded from responsibility for harms caused by their sites, when these harms are caused not only by content but by the promotion and concentration of content through algorithms. As a community, we expect corporations and industries to be responsible for not only their products, but also for any by-products that they produce. Tort law developed in the nineteenth century in response to the early harms of industrialisation. Product liability laws developed in the twentieth century to mandate liability in most jurisdictions for personal injury caused by negligently manufactured products. The tort case of *Donoghue v Stevenson* reformulated the responsibility owed by one person to another in civil society.[46] The case introduces the figure of the neighbour, from which, as Jane Stapleton eloquently describes, a "golden thread" of vulnerability runs into the present.[47] The case is also important to social understandings of responsibility. Tort law has been thought of as a kind of moral praxis—as the real-world, textured complement to philosophical abstraction, and as a key site in which civil norms are produced. As Peter Cane argues, "outside of the law, there are relatively few norm-enforcing institutions in civil society".[48] The birth of negligence law as we know it also involved the demystification of capitalist processes, and the re-imagination of relationships in the market economy. The law concerning the manufacture of objects up until the case of *Donoghue v Stevenson* held that liability for defective or dangerous products could only be based in contract, with the exceptions of fraud and products that were dangerous-in-themselves.[49] In the late twentieth century, governments regulated industry for harms caused to the environment and by their packaging. Applying the same logic, social media companies should be held responsible for the harms caused by their negligence and their failures to produce both the users of the internet and workers who are employed to make their products safer.

The uploading of the Christchurch terrorist video became a watershed moment, turning people's attention again to the question of whether, as Tristan Harris, a former Google employee turned critic of the tech industry, says, Facebook is an "uncontrollable digital Frankenstein", or whether we should be looking at introducing government regulation (state-based or international) or new legal approaches.[50] The level of moderation provided by social-media corporations is inadequate, but the moderation that *does* take place is outsourced to third parties, and this moderation takes place under conditions that are unethical and that disproportionately affect less powerful people. The work of moderation involves technicians looking at images and text depicting violence, pornography and death to judge whether they violate the policies of Facebook, Twitter, Instagram or YouTube—everything from online bullying to child porn and photos of dead bodies. AI is also used. In Facebook's fifth Community Standards Enforcement Report, released in May 2020,[51] it states that AI is used to detect 90% of hate speech. But AI does not always recognise harmful content in video images and over-polices sub-cultures such as the LGBTQI community. Human moderators are needed to detect political and historical context, as well as nuance and wordplay. A number of third-party-companies contract with companies such as Facebook and employ moderators. This work will inevitably be difficult, but the problem with the way in which it is being done to date is that it is harmful to workers. Currently, moderators are inundated with images after little training, and are compelled to view high rates of text/images with low rates of evaluation error. They are not provided with adequate after-care. Altogether, this has led to an unsafe work environment. In 2018, a group of US moderators sued Facebook on this basis. The moderators claimed that reviewing violent, graphic images led to them developing post-traumatic stress disorder. In May 2020, Facebook agreed to pay $52m in settlement of a class action by a group of content moderators as compensation for mental health harm developed as a result of their work as moderators. Facebook also agreed to start using new tools intended to reduce the harm of moderators viewing harmful content.[52] However, as recently as January 2020, Accenture, one of the third-party contractors that hires moderators, started to ask workers to sign liability waivers in which workers acknowledge that their job could lead to PTSD, thereby laying the groundwork to outsource responsibility for harm in the future.[53] A similar class action has been filed in Ireland for plaintiffs across Europe.[54]

Thinking through the bias problem

The question of technology and bias can be primarily addressed along two lines: (1) that institutional settings need to be radically changed to reflect diverse populations, so that the people making tech are diverse; and (2) that bias can and should be designed out of tech. We consider both these below, through the lens of critical legal theory.

Institutions and workplaces

Timinit Gebru, who co-founded Black in AI, a group set up to encourage people of colour to join the artificial intelligence field, is a prominent critic of the lack of diversity in tech companies and the biases present in metrics of employability such as IQ tests and standardised testing. Only 2.5% of Google's workforce is Black, whereas at Facebook and Microsoft the rates are 4% each; only 15% of AI research staff at Facebook are women, and only 10% at Google, while there is no data on trans or other gender minority workers.[55] The "diversity problem" is intersectional, and based on structural as well as historical continuities. It affects workplace cultures and also how tech products are made, who they're designed to benefit, and where their harms fall. The diversity problems in tech workplaces not only reflect but produce the bias in tech. In fact, we might think tech itself "as systems of discrimination".[56]

The AI Now Institute's *Discriminating Systems: Gender, Race, and Power in AI* report is the first report in a series examining the 'diversity crisis' in the artificial intelligence sector. Key findings of the Report include that: (1) The overwhelming focus on 'women in tech' is likely to privilege white women; (2) Fixing the 'pipeline' (ie. the flow of diverse job candidates from school into industry) won't fix AI's diversity problem because it does not address deeper issues such as sexual harassment, unfair compensation, tokenization, and workplace culture; and (3) The use of AI systems for the classification, detection, and prediction of race and gender is in urgent need of re-evaluation.[57] The first two findings critique the dominant reforms offered by AI bias scholarship, that the 'pipeline' approach to increase workplace diversity will provide a solution. West et al write, on the third point, "the histories of 'race science' are a grim reminder that race and gender classification based on appearance is scientifically flawed and easily abused".[58] There is a real danger that technology is reinscribing the idea that race has certain phenotypical features, that race is "real" rather than ideological

AI Now makes recommendations for reform along three main lines. First, that there should be greater transparency in institutions regarding compensation levels along race and gender lines, publication of harassment and discrimination claims and around hiring processes.[59] Second, there should be more practices that improve workplace diversity—recruitment that goes beyond elite universities to include pathways for contractors and temps to become full-time, targeted increases in minorities becoming leaders, tying executive incentive structures to the retention of minorities; and ensuring diversity in AI research.[60] Finally, there should be greater policies for addressing bias and discrimination. There should be rigorous testing including pre-release trials of AI design projects, independent auditing, and greater expertise beyond technical expertise to gain insight into bias.[61]

We also need to think more deeply about the history and investments of institutions, and law's complicity in these processes. Sara Ahmed has looked at the ways in which power has played out in institutions under the banner of "diversity". In her book *On Being Included*, Ahmed argues that diversity is an ordinary, even unremarkable, feature of institutional life. Yet diversity practitioners often experience institutions as resistant to their work, as captured through their use of the metaphor of the "brick wall". Ahmed explores the gap of this paradox, between symbolic commitments to diversity and the experience of those who embody diversity. Commitments to diversity are understood as "non-performatives" that do not bring about what they name. Ahmed shows how racism can be in fact abetted and obscured by the institutionalization of diversity, while diversity is used as evidence that institutions do not have a problem with racism.

Designing out bias

Institutional diversity is crucial but it is not possible to hire a group of people that is completely representative; nor, as Ahmed points out, is it possible to completely guard against institutions using diversity cynically. Another arm of work with bias is to design out problems—to develop proactive plans that anticipate and preempt problems for specific groups. In the case of dirty policing, for example, Richardson et al highlight that reform must effectively 'break' the confirmation feedback loop at the 'dirty policing' stage. As Richardson argues, merely identifying unlawful and biased practices is not enough… the data collection, analysis, and use of these practices must be reformed".[62]

Critical theorists bring to this the additional insights that we need to think through the specific histories and structures of oppression. The concept of intersectionality is crucial to understanding how bias plays out in AI. Speaking at thirtieth anniversary on intersectionality in February 2020, Krenshaw stated:

> It's basically a lens, a prism, for seeing the way in which various forms of inequality often operate together and exacerbate each other. We tend to talk about race inequality as separate from inequality based on gender, class, sexuality or immigrant status. What's often missing is how some people are subject to all of these, and the experience is not just the sum of its parts.[63]

Critical theory suggests it will not be enough to push more diverse institutions and data sets—that instead, we need to look at the relevant concepts and the material histories. We should question the nature of the goals that AI is said to serve. Some of these approaches are extra-legal but there is a role for law in overseeing the design and institutional setting of tech—in making sure each is not discriminatory. There are precedents for how this might be achieved in anti-discrimination law, but we suggest that law and legal thinkers should take both a wider and deeper role in educating the tech sector regarding the history of both institutional and design settings—and how these contexts actively produce tech.

Notes

1. Carmen Niethammer, "AI Bias Could Put Women's Lives At Risk—A Challenge for Regulators" *Forbes*, 2020, https://www.forbes.com/sites/carmenniethammer/2020/03/02/ai-bias-could-put-womens-lives-at-riska-challenge-for-regulators/#3f82d4a0534f Accessed 1 September 2020.
2. John Edwards, "AI Regulation: Has the Time Arrived?" (*Information Week* 2020) https://www.informationweek.com/big-data/ai-machine-learning/ai-regulation-has-the-time-arrived/a/d-id/1337099 Accessed 1 September 2020.

3. The European Commission (2020) *White Paper on Artificial Intelligence: a European approach to excellence and trust* (released on Feb. 19, 2020).
4. The European Commission (2020) *White Paper on Artificial Intelligence: a European approach to excellence and trust* (released on Feb. 19, 2020), p19 para 3.
5. Marcia Wendorf, "Facial Recognition Technology is Being Used to Find Missing Children," *Interesting Engineering*, 2020, https://interestingengineering.com/facial-recognition-technology-is-being-used-to-find-missing-children Accessed 1 September 2020.
6. Mark Farragher, "Five amazing things AI is doing that will blow your mind" (Medium 2019). https://medium.com/machinelearningadvantage/five-amazing-things-ai-is-doing-that-will-blow-your-mind-5e4e96401797 Accessed 1 September 2020.
7. *Science Daily* 2019 https://www.sciencedaily.com/releases/2019/05/190506150126.htm Accessed 1 September 2020.
8. Algorithmic Justice League https://www.ajl.org Accessed 1 September 2020.
9. Steve Lohr, "Facial Recognition is Accurate, if You're a White Guy," *The New York Times* 2018 https://www.nytimes.com/2018/02/09/technology/facial-recognition-race-artificial-intelligence.html Accessed 1 September 2020.
10. Poppy Noor, "Can we trust AI not to further embed racial bias and prejudice?" *The BMJ*, 12 February 2020. Accessed 1 September 2020, 1.
11. Sarah Perez, "Microsoft silences its new AI bot Tay, after Twitter users teach it racism" Tech Crunch, March 25 2016, https://techcrunch.com/2016/03/24/microsoft-silences-its-new-a-i-bot-tay-after-twitter-users-teach-it-racism/ Accessed 1 September 2020.
12. Timnit Gebru, "Race and Gender," *Oxford Handbook on AI Ethics* (Oxford University Press 2019), 12.
13. Ibid, 2.
14. Timnit Gebru, note 12, 9.
15. Ibid.
16. Rashida Richardson, Jason Schultz and Kate Crawford, "Dirty Data, Bad Predictions: How civil rights violations impact police data, predictive policing systems, and justice," (2019) 94:15 pp.15–55, 21.
17. Ibid, 25.
18. Ruth Wilson Gilmore, *Golden Gulag: Prisons, Surplus, Crisis, and Opposition in Globalizing California* (Berkeley: University of California Press 2006), 28.
19. Thalia Anthony, "FactCheck Q&A: are Indigenous Australians the most incarcerated people on Earth?" *The Conversation* 6 June 2017 https://theconversation.com/factcheck-qanda-are-indigenous-australians-the-most- incarcerated-people-on-earth-78528 Accessed 1 September 2020.
20. Susan Chenery, "Women In Prison: It took six months of anguish to get my child," *The Guardian*. 24 February 2019.https://www.theguardian.com/australia-news/2019/feb/24/women-in-prison-it-took-six-months-of-anguish-to-get-my-child. Accessed 1 September 2020.
21. Noor, n10, p1.
22. Noor, n10, p1–2.

23. "Gartner says nearly half of CIOs are planning to deploy artificial intelligence," Press Release, February 13, 2018, https://www.gartner.com/en/newsroom/press-releases/2018-02-13-gartner-says-nearly-half-of-cios-are-planning-to-deploy-artificial-intelligence Accessed 1 September 2020.
24. Zoe Quinn. *Crash Override: How Gamergate (Nearly) Destroyed My Life, and How We Can Win the Fight Against Online Hate* (PublicAffairs 2017). 48.
25. Ibid.
26. Danielle Keats Citron, *Hate Crimes in Cyberspace* (Harvard University Press 2014). 19.
27. Ginger Gorman, *Troll Hunting: Inside the World of Online Hate and its Human Fallout* (Hardie Grant 2019). 16.
28. Emma A Jane (2018) "Gendered Cyberhate as Workplace Harassment and Economic Vandalism," *Feminist Media Studies* 18(4). 2, 15.
29. Emma A Jane, *Misogyny Online: A Short (and Brutish) History* (SAGE Publications 2017). 56.
30. Amnesty International. "Chapter 2—Triggers of Violence and Abuse Against Women on Twitter" in *Toxic Twitter – Women's Experiences of Violence and Abuse on Twitter.* https://www.amnesty.org/en/latest/research/2018/03/online-violence-against-women-chapter-2/#topanchor. Accessed 1 September 2020.
31. Tom Swann, "Trolls and polls—surveying economic costs of cyberhate," *The Australia Institute*, 28 January 2019. https://www.tai.org.au/content/online-harassment-and-cyberhate-costs-australians-37b. Accessed 1 September 2020.
32. Maeve Duggan, "Part 1: Experiencing Online Harassment," Pew Research Center Internet & Technology, October 22, 2014. http://www.pewinternet.org/2014/10/22/part-1-experiencing-online-harassment/. Accessed 1 September 2020.
33. Amnesty International. "Chapter 2—Triggers of Violence and Abuse Against Women on Twitter" in *Toxic Twitter – Women's Experiences of Violence and Abuse on Twitter.* https://www.amnesty.org/en/latest/research/2018/03/online-violence-against-women-chapter-2/#topanchor. Accessed 1 September 2020.
34. https://www.facebook.com/communitystandards/. Accessed 1 September 2020.
35. Tony Romm, "Facebook reports it took action against tens of millions of posts for breaking rules on hate speech, harassment and child exploitation," *The Washington Post*, 14 November 2019, https://www.washingtonpost.com/technology/2019/11/13/facebook-reports-it-took-action-against-tens-millions-posts-breaking-rules-hate-speech-harassment-child-exploitation/. 1 December 2020.
36. Ibid.
37. Thomas Brown, "Social Media and Online Platforms as Publishers," Research Briefing, House of Lords Library, January 8, 2018. https://researchbriefings.parliament.uk/ResearchBriefing/Summary/LLN-2018-0003#fullreport. Accessed 1 September 2020.
38. Amnesty International. "Chapter 7 – Human Rights Violations" in *Toxic Twitter – Women's Experiences of Violence and Abuse on Twitter.* https://www.amnesty.org/en/latest/research/2018/03/online-violence-against-women-chapter-2/#topanchor. Accessed 1 September 2020.

39. Emily Bell, "Technology Company? Publisher? The Lines Can No Longer Be Blurred," *The Guardian*, April 2, 2017. https://www.theguardian.com/media/2017/apr/02/facebook-google-youtube-inappropriate-advertising-fake-news. Accessed 1 September 2020.
40. Lee Rainie, Janna Anderson and Jonathan Albright, "The Future of Free Speech, Trolls, Anonymity and Fake News Online. Pew Research Center," Pew Research Centre, March 2017, 11. http://assets.pewresearch.org/wp-content/uploads/sites/14/2017/03/28162208/PI_2017.03.29_Social-Climate_FINAL.pdf. Accessed 1 September 2020.
41. The United States Department of Justice, "Department of Justice's Review of Section 230 of the Communications Decency Act of 1996," https://www.justice.gov/ag/department-justice-s-review-section-230-communications-decency-act-1996. Accessed 1 September 2020.
42. Nitasha Tiku, "How a Controversial New Sex-Trafficking Law Will Change the Web," *Wired*, March 22, 2018. https://www.wired.com/story/how-a-controversial-new-sex-trafficking-law-will-change-the-web/. Accessed 1 September 2020.
43. Colin Lecher, "Sen. Ron Wyden on Breaking Up Facebook, Net Neutrality, and the Law that Built the Internet," *The Verge*, July 24, 2018. https://www.theverge.com/2018/7/24/17606974/oregon-senator-ron-wyden-interview-internetsection-230-net-neutrality. Accessed 1 September 2020.
44. Ibid.
45. Whitney Phillips, *This is Why We Can't Have Nice Things*: Mapping the Relationship between Online Trolling and Mainstream Culture (MIT Press 2016).
46. Honni van Rijswijk, "Neighbourly Injuries: Proximity in Tort Law and Virginia Woolf's Theory of Suffering," (2012) *Feminist Legal Studies* 20(1): 39–60.
47. Jane Stapleton, "The Golden Thread at the Heart of Tort Law: Protection of the Vulnerable," in Peter Cane (ed.), *Centenary Essays for the High Court of Australia* (LexisNexis Butterworths 2004) pp. 242–56, p259.
48. Peter Cane, *Responsibility in Law and Morality* (Hart 2002). 25.
49. Donoghue v Stevenson 1932 A.C. 562.
50. Tristan Harris, Twitter, March 18, 2019
51. https://transparency.facebook.com/community-standards-enforcement
52. "Facebook to pay $52m to content moderators over PTSD," *BBC News* 12 May 2020, https://www.bbc.com/news/technology-52642633. Accessed 1 September 2020.
53. "Facebook and YouTube moderators sign PTSD disclosure," *BBC News* 25 January 2020, https://www.bbc.com/news/technology-51245616. Accessed 1 September 2020.
54. Jennifer O'Connell, "Moderators to take Facebook to court for 'psychological trauma'," *The Irish Times*, 7 September 2019, https://www.irishtimes.com/news/crime-and-law/moderators-to-take-facebook-to-court-for-psychological-trauma-1.4010391. Accessed 1 September 2020.
55. Tom Simonite (2018) AI is the future—but where are the women? *WIRED*. https://www.wired.com/story/artificial-intelligence-researchers-gender-imbalance/ Accessed 1 September 2020.

56. Sarah West, Meredith Whittaker and Kate Crawford (2019). *Discriminating Systems: Gender, Race and Power in AI*, AI Now Institute https://ainowinstitute.org/ discriminatingsystems.html Accessed 1 September 2020.
57. Ibid. Listed on page 3.
58. Sarah West, Meredith Whittaker and Kate Crawford, n 55, p3.
59. Sarah West, Meredith Whittaker and Kate Crawford, n 55, 4.
60. Ibid.
61. Ibid.
62. Richardson et al, n 16, 37.
63. Katy Steinmetz, "She Coined the Term 'Intersectionality' Over 30 Years Ago. Here's What it Means to Her Today," *Time*, 20 February 2020, https://time.com/5786710/kimberle-crenshaw-intersectionality/. Accessed 1 September 2020. Kimberlee Crenshaw, (1991) "Mapping the margins: Intersectionality, identity politics, and violence against women of color." *Stanford Law Review*. 43: 1241–99.

Index

Note: Page numbers followed by 'n' refer to notes.

Printed in the United States
by Baker & Taylor Publisher Services